THE DYNAMIC STRUCTURE
OF THE DEEP EARTH

THE DYNAMIC STRUCTURE
OF THE DEEP EARTH

An Interdisciplinary Approach

Shun-ichiro Karato

PRINCETON UNIVERSITY PRESS

Princeton and Oxford

LIBRARY OF CONGRESS CATALOGING-IN-PUBLICATION DATA
Karato, Shun-ichiro, 1949–
    [Reoroji to chikyu kagaku. English]
    The dynamic structure of the deep earth : an interdisciplinary approach /
Shun-ichiro Karato.
        p. cm.
    Includes bibliographical references and index.
    ISBN 0-691-09511-6 (acid-free paper)
    1. Earth—Internal structure.   2. Earth sciences.   3. Rheology.   I. Title.
QE509 .K3713 2003
    551.1′1—dc21   2002029298

British Library Cataloging-in-Publication Data is available

This book has been composed in Sabon and Bluejack

Printed on acid-free paper.∞

www.pupress.princeton.edu

Printed in the United States of America

10 9 8 7 6 5 4 3 2 1

CONTENTS

PREFACE

The last thirty years or so have witnessed a marked change in our view of how this planet may work. In the late 1960s, observations on the ocean floor revolutionized our view of solid Earth from a static to a dynamic one: we found that solid Earth is not a dead body, but that it has been deforming along localized regions, which causes a range of geological activities. This dynamic view of solid Earth, *plate tectonics,* was developed very quickly; by the mid-1970s, the motion of the surface layer (the lithosphere) of Earth had been well documented. However, little was known about the processes in the deep interior of Earth that are associated with the motion of lithospheric plates. In the mid-1970s, we knew that the lithospheric plates created at mid-ocean ridges sink into the mantle at ocean trenches, but we had little idea about how they penetrate deep into the mantle. We knew nothing about the central portion, the inner core, except that it is made largely of solid iron.

However, such a situation has changed during the last twenty years through the advances in three major branches of solid Earth sciences: seismology, high-pressure mineral physics, and modeling through numerical simulations. Through seismology, surprisingly complicated structures and dynamic processes in Earth's deep interior have been found. We now know that not only its surface but its deep interior are under dynamic action. *Seismic tomography* shows that the lithospheric plates penetrating into the mantle encounter great resistance forces in the mid-mantle depth (the transition zone) and are deformed, which is sometimes associated with deep earthquakes, and finally penetrate into the lower mantle, at least in the recent past. Mantle convection results in chemical differentiation, forming continents at the surface as well as "anti-continents" at the bottom of the mantle: there is strong evidence for the chemically heterogeneous regions in the deep mantle. The center of Earth, the inner core, is highly anisotropic and is probably rotating faster than the mantle, suggesting that this central portion is also under dynamic action. *Mineral physics* studies through high-pressure experiments and quantum mechanical calculations have shown that the physical properties of materials in the deep Earth are totally different from those in the shallow regions.

Most of Earth's materials can dissolve a large amount of hydrogen (water) under deep Earth conditions, which leads to a dramatic reduction in viscosity. The maximum amount of water that can be dissolved in mantle minerals far exceeds the total amount of sea water. The asthenosphere, a weak layer beneath the lithospheric plate, is caused by the high water content and not by the result of partial melting. Simple materials such as Fe and MgO become highly anisotropic in deep Earth conditions, and seismic waves propagating through them send messages about the flow pattern in regions containing these materials. *Numerical modeling* using high-speed computers has shown a rich variety of interactions between convection currents and phase transformations. Successful modeling of the geodynamo has been made using powerful computers, which suggests the importance of the inner-outer core structure for the generation of the geomagnetic field, suggesting a close link between the thermal evolution of a planet and its geomagnetic field.

One of the key ingredients in this progress is our better understanding of *materials properties under deep Earth conditions*. We can infer the convection pattern from seismological data only when we know how convection modifies seismic wave propagation. Interactions of convection currents with deep mantle layers can be understood only when we know the nature of changes in properties in these regions associated with phase transformations. In short, the full spectrum of a rich variety of deep Earth dynamics can be appreciated only through the understanding of the properties of Earth materials. However, such an aspect is not always appreciated in the geophysical community, nor, consequently, in many textbooks on geophysics. *Geophysics* had traditionally been a branch of *applied mathematics* rather than a branch of natural sciences, and Earth has often been treated only as a spherical object to which some sophisticated mathematical treatment may be applied. There are excellent textbooks on materials science as applied to Earth materials, but they do not provide any in-depth discussions of geodynamic problems that are closely related to materials properties. Recent advancements just summarized above show, however, that a more interdisciplinary approach is needed in solid Earth geodynamics to get a better understanding of the dynamics and evolution of this planet. The purpose of this book is to illustrate how such an interdisciplinary approach may be made using several topics of geodynamics that I have been involved in one way or another. As such, this book is not intended to be a textbook from which a reader can get systematic knowledge of some branches of Earth sciences. Consequently, for the most part this book discusses highly controversial issues in solid Earth geodynamics, and in many places it was necessary to present my own view, although I have tried to provide a balanced picture of the subject. I have not avoided showing the zigzag ways of working scientists who are trying to reveal the secrets of nature; thus, I have taken a risk of making this book more difficult to understand. However, I believe that such a style may indeed be appropriate in showing the excitement of science to a general audience. The intended readerships include upper-level undergraduate students or graduate students in Earth sciences, as well as in physics and materials science.

This present volume is a translation of a book that was published in Japanese in 2000 by the University of Tokyo Press. The translation has been made by Tomoko Korenaga, although significant changes have been made to update the content. In this new version, many of the technical details are given in "boxes" to make a smooth flow of the text without the reader being bothered by the details. Chapter 4 has been almost completely rewritten and chapters 6 and 7 of the Japanese version are combined to a single chapter in this new version. I thank the editors, Mika Komatsu and Kei Shimizu at the University of Tokyo Press and Joe Wisnovsky at Princeton University Press, for their help. Yoshibumi Tomoda, Daisuke Yamazaki, and Phil Skemer read the drafts of this book at various stages. Their comments significantly improved its presentation. Haemyeong Jung helped prepare some of the figures. Finally, I thank my wife Yoko. Without her understanding and support, this book would not have been completed.

Shun-ichiro Karato

THE DYNAMIC STRUCTURE
OF THE DEEP EARTH

ONE • THE STRUCTURE OF EARTH AND ITS CONSTITUENTS

1-1. EARTH'S INTERIOR: RADIAL STRUCTURE, CHEMICAL COMPOSITION, AND PHASE TRANSFORMATIONS

Inferring Earth's composition is a prerequisite to understanding its evolution and dynamics as well as those of planets like it. One might think that the composition of Earth can be easily inferred from the rocks that we can see on Earth's surface. However, it immediately becomes obvious that these rocks cannot be the major constituent of Earth's interior because the densities of typical rocks on Earth's surface, such as granite or basalt, are so small, even if the effects of compression on density are taken into account. Therefore, materials in the deep Earth (and most other planets) are different from those on the surface. What materials are there, and how do we infer the composition of the deep interior of Earth (and other planets)? You may want to drill into Earth, but the deepest drilled hole in the world is in the Kola peninsula in Russia, which is only ~ 12 km deep (remember that Earth's radius is 6,370 km). Although some kinds of volcanoes bring materials from the deep mantle, this sampling is usually limited to ~ 200 km deep. Therefore, our inference of Earth's internal structure must be based largely on indirect information. In this connection, both geochemical and geophysical observations are particularly relevant. In *geochemistry*, the scientist measures the chemical composition of various materials, then uses the chemical rules that govern the distribution of various chemical elements to infer the chemical composition of Earth. In *geophysics*, one measures physical properties such as seismic wave velocities and density, and infers the composition and structure of Earth's interior based on the physical principles that control the variation of physical properties with thermodynamic conditions (pressure and temperature). In this chapter, I will first summarize these basic observations, and then explain several models of Earth that have been proposed on the basis of geo-

1

chemical and/or geophysical lines of inference. Although the inference of composition through such a process is indirect and not unique, some aspects of Earth's interior are now well understood. However, a number of problems remain controversial, including the nature of chemical heterogeneity in Earth's mantle.

1-1-1. Geochemical Models

An obvious starting point for inferring the chemical composition of Earth is the composition of rocks that we can collect on Earth's surface. We have a large data set of the composition of these rocks. Rocks from the shallow regions, the crust, are typically basalt or granite (rocks made mostly of quartz, pyroxenes, and plagioclase) or rocks that have been modified from these rocks by later chemical reactions (metamorphism). These rocks have a high silica content and small densities. Occasionally, we find denser rocks, peridotite or eclogite (rocks made mostly of olivine, pyroxenes, and spinels or garnets), which contain a smaller amount of silica and higher densities than basalt or granite. These rocks are obviously the candidates of mantle materials. However, petrological and geochemical studies show that the rocks that we can see on Earth's surface may not be representative of the bulk of Earth's mantle. It is important to note that Earth's shallow region has undergone extensive chemical differentiation through partial melting (melting of only some components) and the composition of each layer in the shallow regions is likely to reflect these differentiation processes. Consequently, the deeper portions of Earth's mantle likely have a composition different from that of rocks that we can collect on Earth's surface. One needs a theory for chemical differentiation in Earth to infer the chemical composition of Earth from the composition of rocks in near surface regions.

Ted Ringwood, at Australian National University (ANU) at Canberra, Australia, was a leader of the study of Earth's interior through a geochemical, petrological approach. He was perhaps the greatest Earth scientist from Australia. He made a number of fundamental contributions to knowledge of the structure and evolution of Earth and other terrestrial (Earth-like) planets. After obtaining a doctorate in geology at Melbourne in 1956, he studied with Francis Birch at Harvard. He returned to Australia in 1959 to take a position at the newly formed institution at ANU, and remained there until his premature death in 1993. Most of Ringwood's predictions or models of Earth's structures and evolution were based on simple principles, but he had an ingenious sense of synthesis, and most of his models have survived the test of time. Ringwood proposed a model of Earth's chemical composition based on a theory of chemical differentiation in Earth (1975). The starting point of his model is the notion that the formation of basalt through partial melting of mantle materials is the most important volcanic activity on Earth. The mantle materials must then be able to produce basalt (more precisely, mid-ocean ridge basalts [MORB]) by partial melting. From this line of argument, Ringwood proposed a hypothetical rock called *pyrolite*, which produces mid-ocean ridge basalt by partial melting, and he proposed that the majority of the mantle must be composed of pyrolite. Pyrolite is rich in magnesium and iron, similar to rocks brought from the mantle, but its silica concentration is slightly higher than typical mantle samples and it has more calcium, aluminum, and other elements. In this model, typical mantle samples on the surface are interpreted as residue from the partial melting of pyrolite. Ringwood suggested that the majority of the mantle (including the upper and lower mantle) is composed of materials with a chemical composition similar to pyrolite. In this hypothesis, the depth variation of density and the elastic properties in the mantle must be explained as a result of phase transitions, compression, and thermal expansion of the same material.

Another model is that Earth has the same chemical composition as the average composition of the solar system. It is generally believed that the Sun and the planets in the solar system were formed as a result of the collapse of a putative primitive solar nebula. Therefore, the composition of the sun and the other materials of the solar system should be approximately the same. The composition of the outer layer of the sun can be inferred from the analysis of its optical properties (indeed, the element helium, He, was discovered by the analysis of the optical spectrum of the Sun, hence its name [*helios* means "the Sun" in Greek]). Another source of information on the composition of the solar system comes from the composition of meteorites. Meteorites are considered to be fragments of materials that failed to become planets. Among various types of meteorites, carbonaceous chondrite is a unique type, which is made of a mixture of various materials including metallic iron, silicates, and organic materials. The age of this type of meteorite, inferred from the composition of radiogenic isotopes, is the oldest (~ 4.56 billion years) among the ages of various objects in the solar system. Therefore, this type of meteorite is considered to be a remnant of the primitive solar system. The chemical composition of carbonaceous chondrite agrees well with that of the Sun except for the volatile elements. Consequently, the composition of the car-

TABLE 1.1
Chemical Composition of Earth (wt%) (after Ringwood, 1975)

	Continental crust	Upper mantle	Pyrolite model	Chondrite model (1)	Chondrite model (2)
MgO	4.4	36.6	38.1	26.3	38.1
Al_2O_3	15.8	4.6	4.6	2.7	3.9
SiO_2	59.1	45.4	45.1	29.8	43.2
CaO	6.4	3.7	3.1	2.6	3.9
FeO	6.6	8.1	7.9	6.4	9.3
other oxides	7.7	1.4	1.2		5.5
Fe				25.8	
Ni				1.7	
Si				3.5	

Note: In chondrite model (1), the light element in the core is assumed to be Si. Chondrite model (2) is a model of chemical composition of the mantle corresponding to the model of core shown in chondrite model (1).

bonaceous chondrite is considered to be representative of the composition of the solar system. Some scientists consider that Earth has a chemical composition that is similar to that of (carbonaceous) chondrite (except for volatile components). This model is often referred to as the *chondrite model*.

Table 1-1 shows chemical compositions corresponding to the pyrolite model and the chondrite model. One significant difference between these models is the ratio of (Mg + Fe)/Si. In the chondrite model, the amount of silicon in the mantle is greater than that in the pyrolite model. Based on various sources of information, on the other hand, the chemical composition of the upper mantle is estimated to be very similar to pyrolite or have slightly less silicon than pyrolite. Therefore, if Earth has the chondritic chemical composition, the amount of silicon in the deep mantle must be greater than in the shallower mantle. One possibility is a silicon-rich lower-mantle model; some people consider that the lower mantle consists mostly of $(Mg,Fe)SiO_3$. Thus, while the ratio of (Mg + Fe)/Si is constant throughout the entire mantle in the pyrolite model, the chondrite model has a smaller (Mg + Fe)/Si ratio in the lower mantle than in the upper mantle.

More complicated geochemical models have also been proposed. Don Anderson at Caltech (e.g., 1989), Eiji Ohtani at Tohoku University in Japan (1985), and Carl Agee, then at Harvard (1993), proposed mantle models containing several chemically different layers. Their thinking is based on a possible scenario for Earth's formation. Based on the results of Apollo missions, many Earth scientists believe that there was an extremely voluminous melting event due to high temperature caused by energy released via high-velocity collisions that occurred during the formation of the planets. This putative extensively molten region is called the *magma ocean*. They proposed that upon cooling of magma ocean, various minerals would have been solidified and would have sunk or floated to form chemical layering.

I must emphasize that these models have many uncertainties; their role is not to give some definitive picture of Earth structure, but to provide testable hypotheses. This point is particularly important in relation to *geochemical* models. Some rules of geochemistry are very well established and nearly independent of physical conditions. A good example is the rule of the radioactive decay of elements, which causes a temporal variation of isotope compositions. Also, the rule of partitioning some elements between liquids (magmas) and solids (minerals) is relatively well established. However, the physical processes that govern the distribution of chemically distinct materials in Earth are highly dependent on material properties that depend strongly on physical and chemical conditions, and these results should not be interpreted dogmatically to reach conclusions on chemical composition. For example, it is common to all models that the mantle basically consists of $(Mg,Fe)_2SiO_4$ (olivine and its high-pressure polymorphs) and $(Mg,Fe)SiO_3$ (pyroxene and its high-pressure polymorphs), and there is no doubt that $(Mg,Fe)O$ is richer in the mantle than in the crust. For more detailed issues such as the depth variation of the (Mg + Fe)/Si ratio, however, we cannot draw conclusions only from these models. For example, Earth as a whole did not necessarily have the average chemical composition of the solar system. In the primitive solar system, its chemical composition was likely spatially variable, and it is unlikely that Earth and carbonaceous chondrites were formed from the same part of the primitive solar system. Furthermore, even if there was a magma ocean, it is not obvious that its solidification resulted in chemical layering. Brian Tonks and Jay Melosh (1990), at the University of Arizona at

Tempe, have argued that crystallization from the magma ocean could result in a nearly homogeneous chemical composition because of stirring and mixing by vigorous convection during solidification.

So far we have discussed only major element compositions. The distribution of *trace elements* (elements that occur only in a minor content) is also important in relation to material circulation in Earth's interior. Because trace elements exist only in a small amount by definition, their influence on physical properties such as density is often small. However, since their distribution varies greatly by partial melting (because most of trace elements have different sizes or different electric charges than the major elements do), the distribution of trace elements provides an important clue as to the processes of partial melting and the resultant chemical evolution of Earth. Important trace elements are *incompatible elements,* such as rare earth elements, which concentrate in the melt phase when rocks partially melt. If we examine the concentration of trace elements in igneous rocks such as basalt, we can find the chemical composition of their parental rocks. From these studies, we now know that there are several source regions in the mantle for igneous rocks. In some regions, rocks there do not contain very much of these incompatible elements. We infer that these regions have undergone extensive chemical differentiation by partial melting that has removed most of the incompatible elements. In other regions, in contrast, rocks contain large amounts of incompatible elements. These regions are inferred to have undergone a lesser degree of partial melting, or a large amount of materials enriched in these elements (sediments, crustal rocks, etc.) have been added in these regions. The differences in chemical composition lead to different *isotope ratios* because different elements have different radioactive decay schemes. In this way, we can infer the chemical *evolution* of different regions in Earth. It is now agreed that there are at least two chemically distinct regions in Earth that have been isolated from each other more than one billion years (Hofmann 1997). The distribution of these source regions is controversial and is the subject of active research (see chap. 4 for detail).

Hydrogen occupies a unique position among the trace elements. In most of the geochemical literature, hydrogen is not treated as a *trace* element, but its quantity in normal minerals is low, and hydrogen is dissolved in melts more than in minerals. Therefore, hydrogen behaves like an *incompatible element.* Like other trace elements, its influence on density is small, but its influence on some physical properties is large. In particular, hydrogen affects melting behavior (melting temperature and the composition of molten materials) and significantly reduces the resistance of materials against plastic flow. Through its effects on plastic properties, hydrogen can also change the nature of seismic wave propagation. I will discuss these issues in some detail in chapter 2.

1-1-2. Geophysical Models

Any model for the interior of Earth (or other planets) must be consistent not only with geochemical observations, but also with geophysical observations. Important geophysical observations include density and elastic properties. The average density of Earth can be easily calculated from geodetic observations of its size and the total mass. The average density of Earth is estimated to be 5,515 kg/m^3. The density of a planet provides an important constraint on its chemical composition. In fact, the average density is almost the only clue to estimate the composition of planets other than Earth, and it is actually sufficient to make a rough guess as to their composition. For example, the density of silicates (minerals consisting mostly of silicon, oxygen, and other elements, such as quartz and olivine) is 2,600–3,400 kg/m^3, and the density of iron (more precisely, iron-nickel alloy) is 7,800 kg/m^3. From these values, we can reach an important conclusion, that Earth consists mainly of silicates and iron. Similarly, we can conclude that the Moon (its density is 3,344 kg/m^3) consists mainly of silicates, and that Ganymede, one of Jupiter's satellites (its density is 1,936 kg/m^3), is composed of silicates and ice. Although we did not consider density variations due to compression and thermal expansion in Earth's deep interior, we can ignore them at this level of discussion. Density variations by pressure and temperature are at most 10–20%, and density differences due to chemical compositions are far more significant.

The moment of inertia is also an important parameter that can be determined by geodetic observation to constrain the internal structure of planets. If a mass M is located at a distance R from an axis of rotation, then the moment of inertia with respect to this rotational axis is MR^2. For a given mass, the moment of inertia is large if the mass is located far from the axis of rotation. Therefore, the moment of inertia depends on the mass distribution within a body, and is small if the mass is concentrated toward the center of a planet. The ratio $C \equiv$ (moment of inertia)/(total mass) × (radius)^2 is a nondimensional number that depends on how mass is concentrated toward the center of a planet. If the mass distribution is uniform, this ratio is 0.4, and if mass is completely concentrated at the center, it is 0.

The estimation of the *moment of inertia* is, however, not so straightforward. Consider a case in which mass (density) changes with depth but its distribution is spherically symmetrical. In this case, we can tell from Gauss's theorem that gravity outside a planet is the same as that in the case of all mass concentrated at the center. Therefore, the depth variation of mass cannot be determined by the measurement of gravity outside a planet. However, in a real world, mass distribution in a planet shows a slight deviation from spherical symmetry due to deformation by the effect of rotation (centrifugal force) and tidal force. Because of this, the moment of inertia can be determined using solely the observations outside a planet. If mass distribution differs slightly from spherical symmetry, objects rotating around a planet (including artificial satellites) are affected not only by the central force but also by the torque. As a result, the orbit of a satellite is not fixed at a perfect elliptical orbit, but the orbital plane moves slowly around the equator of a planet. By measuring this movement of the orbital plane, we can estimate how mass distribution in a planet deviates from the spherical symmetry. At the same time, this torque affects the motion of the planet itself and causes *precession*. Precession is significant when the cause of torque (the mass of other planets) is large. Thus, while the effect of artificial satellites on the precession of a planet can be ignored, the existence of large celestial bodies close to a planet in consideration has a significant influence on precession. The period of precession depends on the magnitude of torque applied to a planet and its moment of inertia. Once the magnitude of torque is estimated from the orbit of the satellite, therefore, the moment of inertia can be calculated by the period of precession. In the case of Earth, it has been found that $C = 0.3308$. From this, we can infer that the mass of Earth is concentrated toward the center; namely, the existence of a heavy core is implied. Precession has not been observed for most of the other planets. In these cases, by measuring deformation due to centrifugal and tidal forces, we can calculate the moment of inertia. Table 1-2 summarizes basic geodetic data on Earth.

In the case of Earth, earthquakes occur frequently, and this allows us to obtain detailed information about the elastic properties and the density of its interior. Earthquakes, which are nothing but a disaster for our daily lives, illuminate the dark interior of Earth for researchers. In fact, almost all of the chapters of this book deal with seismological observations, which give us the most detailed information on Earth's interior. Figure 1-1 shows how seismic waves propagate through Earth. Body waves (seismic waves that propagate through the bulk of Earth), which have been studied since the earliest time of seismology, propagate three-dimensionally from a hypocenter (the location where an earthquake oc-

TABLE 1-2
Basic geodetic data for Earth

equatorial radius	6,378 km
polar radius	6,357 km
flattening	1/298.26
total volume	1.083×10^{21} m^3
total mass, M	5.9737×10^{24} kg
average density	5,515 kg/m^3
moment of inertia around the spin axis, I	8.036×10^{37} kg m^2
$C = I/MR^2$ (R: average radius)	0.3308

curs). Most of Earth's interior is solid, and there are two types of seismic waves (elastic waves): compressional and shear waves. The velocities of these two waves are determined by the elastic properties and density of a material, and they satisfy the following relations:

$$V_p = \sqrt{(K + (\tfrac{4}{3})\mu)/\rho}, \ V_s = \sqrt{\mu/\rho}, \tag{1-1}$$

where V_p and V_s are the velocities of compressional and shear waves (p stands for *primary* and s stands for *secondary*), K is bulk modulus or incompressibility, μ is shear modulus or rigidity, and ρ is density. By knowing the time and place of an earthquake (which can be inferred from the travel times at various stations), the seismic wave velocities of a region through which seismic waves propagate can be determined from the travel times that seismic waves take to arrive at observational points (seismological stations). Seismic waves also reflect and refract at various boundaries. From reflection coefficients, we can obtain the information not only on seismic velocities but also on densities.

Seismic waves that reach to a boundary undergo reflections and refractions to cause a certain type of wave that propagates along the surface. These are called *surface waves*. Analysis of surface wave propagation is more sophisticated than that of body waves. Progress in computer technology and in theoretical treatment has made it possible to analyze in

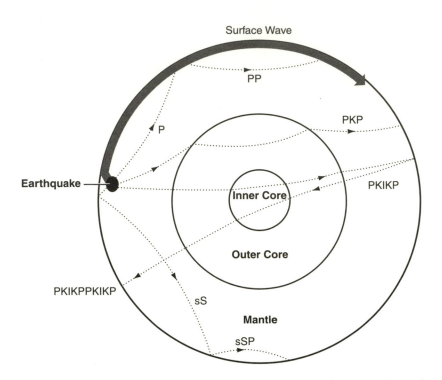

Fig. 1-1. The propagation of seismic waves. The broken lines show the propagation paths of body waves. Body waves are named (PP, SKS, etc.) based on their paths. The thick solid line denotes the propagation path of the surface wave.

detail the seismic wave propagation through complex structures. The surface wave study of Earth's internal structure is especially important for the study of the upper mantle: there is a low seismic velocity zone in the upper mantle (chapter 2), and body waves avoid traveling through the low-velocity zone. The velocity of surface wave varies with wavelength. This is called the dispersion relation. Dispersion occurs because elastic properties vary with depth and because waves with a longer wavelength are sensitive to elastic properties at greater depths. Thus, from the dispersion of surface waves, the structure of Earth can be inferred. One of the advantages of surface wave study is that it provides information on low-velocity zones. Since surface waves propagate two-dimensionally, they can propagate for long distances without much attenuation. Therefore, when a big earthquake occurs, surface waves traveling around Earth many times give us a large amount of data regarding the relatively shallow part of Earth.

Traveling surface waves can cause a shaking of Earth as a whole. This phenomenon is referred to as *free oscillation*. Earth rings just like a big bell; because its *tone* depends on the elastic properties and density of Earth, as happens in the case of a regular bell, we can investigate the interior of Earth by analyzing the *tone* of Earth. The principle is similar to that used in the study of surface waves. Free oscillations have various *modes* (like harmonic overtones for musical instruments), and each mode is sensitive to elastic properties and density at a different depth. The structure of Earth can be determined by comparing observations and models for various mode frequencies. The first observation of Earth's free oscillation was made during the Chilean earthquake in 1960, and free oscillations have been used for the study of Earth's internal structure ever since. One of the advantages of studies using surface waves and free oscillations is that it is not necessary to know the time and location of the earthquake. In body wave studies, the uncertainty of hypocenters (locations of earthquakes) can result in large errors.

Since the study of surface waves and free oscillations uses lower-frequency waves than those of body waves, it is sometimes called *low-frequency seismology*. Although this method can accurately determine large-scale structures, it cannot determine small-scale structures because low-frequency waves have long wavelengths, and hence their propagation is insensitive to small-scale features. It is therefore important to use a range of data, including body waves, surface waves, and free oscillations, to constrain the structure of Earth.

By integrating various kinds of seismic data, Adam Dziewonski, at Harvard, and Don Anderson proposed a standard model for the internal structure of Earth. In this model, a range of seismological observations are included in the analysis, and the corrections for the effects of using different frequencies are made based on the physical model of elastic and non-elastic deformation (box 1-1). This model is called PREM (Preliminary Reference Earth Model) (fig. 1-2) (Dziewonski and Anderson 1981). For their fundamental contributions to the study of the structure of Earth's interior through seismology they received the Crafoord Prize, equivalent to the Nobel Prize, in 1998 from the Swedish Academy. By comparing the densities and elastic constants of various materials, the likely composition of Earth can be inferred from the distribution of density and elastic constants within Earth as they are given by PREM. For a rough estimate of its chemical composition, we can ignore the effects of pressure and temperature on density. The previous conclusion that Earth is composed of silicates and iron can be derived from this kind of ap-

Box 1-1. Anelasticity and Physical Dispersion

One of the important points of PREM is the incorporation of a range of seismological observations in a physically consistent fashion. To a very high degree, the propagation of seismic waves can be understood based on the theory of elastic waves. However, when the details of wave propagation are investigated, deviations from perfect elasticity can be noted. This is due to the fact that seismic waves are low-frequency waves (~ 0.001 to ~ 1 Hz) that propagate through rather hot materials: the temperatures in most portions of Earth exceed half of melting temperature. Anelasticity causes dissipation of energy as heat; consequently, the amplitude of elastic waves decreases with time (and hence distance), leading to seismic wave attenuation. When a material responds to an external force with some energy dissipation, then elastic constants of that material become dependent on frequency. At infinite frequency, there is no time for a viscous element to respond, and there is no effect of anelasticity: elastic constants are the same as the case without any anelasticity. In contrast, at lower frequencies, there is more time for viscous components to affect the response of a material; hence, the elastic constants (seismic wave velocities) decrease with the decrease of frequency. This is called physical dispersion (an example of what is known the *Kramers-Kronig relation* in physics). Its importance was first pointed out by Gueguen and Mercier, in France (1973), and first demonstrated in seismology by Hiroo Kanamori and Don Anderson (1977). Anelasticity therefore provides a link between seismological observations and rheological properties.

proach. When we want to consider elastic properties in addition to density, we cannot ignore the effects of temperature, and especially of pressure, anymore. Elastic constants vary by several times for the range of pressure expected in Earth's interior. I will discuss the behavior of materials under high pressure in the later part (sec. 1-1-3) of this chapter.

Based largely on the studies on seismology and high-pressure mineral physics, our understanding of Earth's structure has made significant progress in the last few decades. First, I will explain the first-order approximation of Earth's structure, then summarize our current understanding of temperature distribution within Earth, which is closely related to geodynamics. Recent progress in the study of the fine structure of Earth

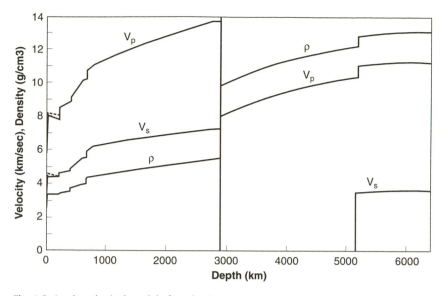

Fig. 1-2. A seismological model of Earth's interior (PREM) (after Dziewonski and Anderson 1981). Vp is the compressional wave velocity; Vs, shear wave velocity; and ρ, density.

(by seismic tomography) has revolutionized the course of research on the dynamics of Earth's deep interior. The progress in geodynamics in connection to seismic tomography will be discussed in chapters 3 and 4.

To construct a first-order model for Earth, we need to know the distributions of pressure and temperature within Earth. The effect of pressure is especially important when estimating chemical composition from density and elastic constants. The effect of temperature is smaller than that of pressure. To a very good approximation, pressure within Earth is in hydrostatic equilibrium. The reason for this approximate hydrostatic balance is that, as will be described later, the viscosity of Earth materials is so small that Earth materials cannot support a large nonhydrostatic stress. The estimated distribution of pressure is shown in figure 1-3. Pressure at the center of Earth is approximately 360 GPa. Although pressure is determined by the basic principle of hydrostatic equilibrium, the distribution of temperature depends on several uncertain factors. Therefore, the temperature distribution, shown in figure 1-3, has large uncertainties. While these uncertainties do not affect the estimation of chemical compositions very much, temperature distribution is important for the dynamics of Earth's interior. I will discuss it in relation to dynamics later in this chapter.

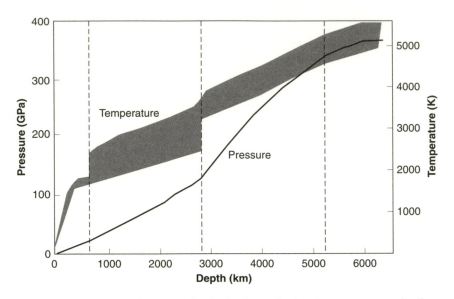

Fig. 1-3. Temperature and pressure distribution in Earth's interior. Temperature distribution greatly depends on the pattern of convection. This figure shows the currently accepted range of temperature distribution.

1-1-3. Earth Models

A. The Layered Structure of Earth: The Zeroth-Order Approximation

There are three chemically distinct layers in Earth that have distinct elastic properties and densities. The crust is a thin and very heterogeneous layer near the surface, which is composed of light silicate minerals such as quartz and feldspar. Crustal thickness varies from place to place; oceanic crust is homogeneous and about 7 km thick, and continental crust is about 30–70 km thick. Beneath the crust, there is a layer composed of denser silicate minerals such as olivine. This region is called the mantle. The mantle continues to the depth of about 2,900 km. Since both compressional and shear waves can propagate through both the crust and the mantle, they must be solid to a large extent. Even if they are molten, the degree of melting should be small in order for shear waves to propagate. At the depth of about 2,900 km, there is a boundary with the largest density jump in Earth. A dense layer beneath this boundary is called the core. Based on its density (and elastic properties), the core is considered to be made mostly of iron.

B. The Layered Structure of Earth: The First-Order Approximation—Phase Transitions

The simplest (zeroth-order) layered structure of Earth described above is due to differences in chemical composition. The chemical composition of each layer is relatively uniform. When we examine each layer closely, however, we can find a variety of layered structures within each layer. The most prominent is the core; while only compressional waves can propagate through the shallower part of the core ($\sim 2,900$ to $\sim 5,150$ km depth), both compressional and shear waves can propagate through the deeper part ($\sim 5,150$ to $\sim 6,370$ km). From this, we can conclude that the shallower part of the core (the outer core) is made mostly of molten iron, whereas the deeper part (the inner core) is made mostly of solid iron. The melting point of iron increases toward the center of Earth because of the increase in pressure, and this probably results in the formation of the solid inner core. This is a typical example of layering due to a phase transformation. As described later, the layered structure of the inner and outer core is thought to play an important role in the generation of geomagnetic fields.

Although not as prominent as above example, the mantle also has important layering. In the depth range of $\sim 410-660$ km, both density and elastic wave velocities increase much faster than in other layers. Francis Birch, at Harvard University, was a father of mineral physics who received his doctorate in physics at Harvard in 1932 under the supervision of Nobel laureate Percy Bridgman, established the first high-pressure mineral physics laboratory at Harvard, in the early 1930s. Birch (1952) showed that it is impossible to explain such a rapid increase in density by the compression of the same material with the same structure alone. He derived this conclusion by comparing the density distribution of Earth estimated from seismic observations with the density distribution at some standard condition. We consider density variation due to the vertical motion of a material with *adiabatic* compression (or expansion) as a standard condition. The adiabatic compression (or expansion) refers to a process in which a material is compressed (or expanded) without an exchange of heat with the surrounding materials. This would occur when a piece of rock is moved vertically in Earth rapidly enough. Such a rapid vertical movement is considered to occur in Earth by vigorous *convection*. The change in density due to adiabatic compression (or expansion) can be inferred from seismic wave velocities. The actual variation of density with depth can also be inferred from seismic wave velocities with the help of other constraints, including the moment of inertia. Therefore, it is possible to compare the depth variation of actual density in Earth with the depth variation in density corresponding to hypothetical adiabatic compression (or expansion). This ratio is called the *Bullen parameter* after a New Zealander seismologist, Keith Bullen:

$$[\text{Bullen parameter}] \equiv \left(\frac{d\rho}{dz}\right) / \left(\frac{d\rho}{dz}\right)_{ad}, \tag{1-2}$$

where $d\rho/dz$ is the density gradient (z is the depth) in a real Earth model and $(d\rho/dz)_{ad}$ is the adiabatic density gradient calculated from seismic wave velocities and gravity (box 1-2). The Bullen parameter is 1 if the density variation in Earth occurs solely due to adiabatic compression (or expansion). A value less than unity indicates that density does not increase much with depth, which may be caused by a large temperature gradient. On the other hand, a value greater than unity corresponds to a density increase larger than expected from the standard condition, which may due to phase transformations. Figure 1-4 shows the distribution of the Bullen parameter. At the shallower part of the upper mantle, it is smaller than unity, suggesting a large temperature gradient (see sec. 1-2). At the deeper part (410–660 km), the parameter significantly exceeds unity. On the basis of this fact, Birth inferred that the mantle minerals undergo phase

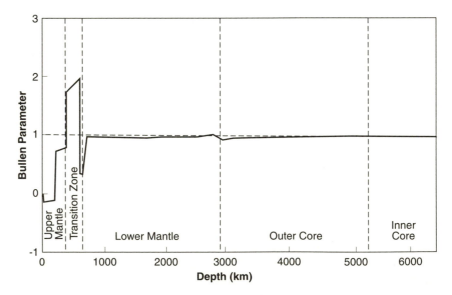

Fig. 1-4. Depth variations of the Bullen parameter (after Dziewonski and Anderson 1981).

Box 1-2. The Bullen Parameter

The Bullen parameter is defined by the ratio of the depth variation of the actual density to the hypothetical depth variation corresponding to adiabatic compression. The adiabatic density gradient, $\left(\frac{d\rho}{dz}\right)_{ad}$, can be determined from seismological observation. Note first that the compression of materials by seismic waves occurs much faster than thermal diffusion. This can be seen by comparing the time-scale for thermal equilibrium with the time-scale of deformation by seismic waves. The time-scale of thermal diffusion is given by $\tau_{th} \sim \lambda^2/\kappa$, where λ is the wave length of seismic waves (~ 10–$1{,}000$ km) and κ is thermal diffusivity ($\sim 10^{-6}\, m^2 s^{-1}$). The time-scale of deformation by seismic wave is $\tau_{def} \sim 1/\omega$, where ω is the frequency of seismic waves. Therefore, $\frac{\tau_{th}}{\tau_{def}} \sim 10^{14-21}$, indicating that the deformation associated with seismic wave propagation is adiabatic. The bulk modulus for adiabatic deformation is defined by

$$K \equiv \left(\frac{dP}{d\log\rho}\right)_{ad}. \tag{B1-2-1}$$

Therefore one gets

$$\left(\frac{d\rho}{dP}\right)_{ad} = \frac{1}{V_\phi^2}, \tag{B1-2-2}$$

where we used the definition $V_\phi^2 \equiv \frac{K_S}{\rho} = V_p^2 - \frac{4}{3}V_s^2$. Now in Earth, pressure is determined by hydrostatic equilibrium, $dP = \rho g\, dz$, thus,

$$\left(\frac{d\rho}{dz}\right)_{ad} = \frac{\rho g}{V_\phi^2}. \tag{B1-2-3}$$

Therefore,

$$(\text{Bullen parameter}) = \frac{\left(\frac{d\rho}{dz}\right)\left(V_p^2 - \frac{4}{3}V_s^2\right)}{\rho g}. \tag{B1-2-4}$$

All of these quantities can be obtained from an Earth model based on seismology.

transformations in the transition zone. This prediction was later confirmed by the experimental work of Ted Ringwood and Lin-Gun Liu in Australia and Syun-iti Akimoto, Naoto Kawai, Mineo Kumazawa, and Eiji Ito in Japan. In addition to an initial quick identification of new

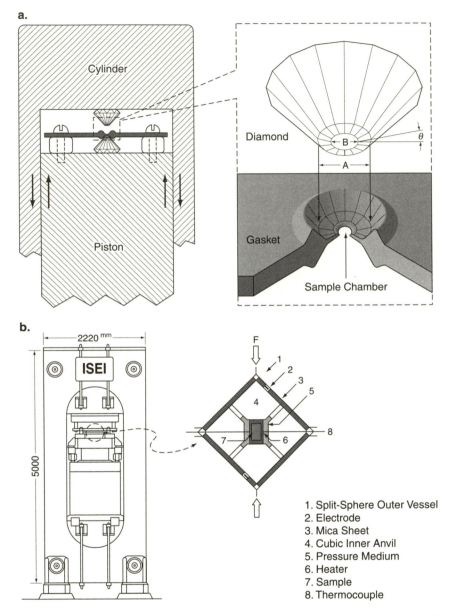

a.

Cylinder

Piston

Diamond

Gasket

Sample Chamber

B

A

θ

b.

2220 mm

5000

ISEI

F

1
2
3
5
4
7 6 8

1. Split-Sphere Outer Vessel
2. Electrode
3. Mica Sheet
4. Cubic Inner Anvil
5. Pressure Medium
6. Heater
7. Sample
8. Thermocouple

Fig. 1-5. (a) A schematic drawing of a diamond anvil cell (DAC). A small sample ($< 10^{-4}$ mm^3) is squeezed between two single crystals of a diamond. A high temperature can also be generated by laser-beam heating. The pressure and temperature that can be generated by a DAC exceed those at the center of Earth. (b) A multi-anvil apparatus designed originally by Naoto Kawai. A relatively large volume of samples ($1-10$ mm^3) can be squeezed. The maximum pressure and temperature conditions achievable with this apparatus are limited to those in the shallow lower mantle. However, since the sample volume is large and the temperature distribution is nearly homogeneous, this type of device is useful for studies of chemical reactions (phase transformations) and plastic properties (from Mao and Hemley 1998).

phases by a laser-heated diamond-anvil cell (DAC; fig. 1-5a), quantitative analyses of phase diagrams using a multianvil apparatus (fig. 1-5b), developed by Kawai and others in Japan, played a major role.

Phase transformations in the mantle transition zone affect the pattern of mantle convection. Recent seismic tomography has suggested that mantle convection seems to change its pattern around these depths (chaps. 3, 4). One of the important themes in the study of mantle dynamics is to understand how convection is affected by phase transformations. At greater depths in the lower mantle and the core, the Bullen parameter is close to 1, suggesting a nearly adiabatic temperature gradient. However, one needs to be careful about temperature gradients. Even if a temperature gradient is twice as large as the adiabatic value, the Bullen parameter varies from 1 to 0.92 at the utmost, which is still consistent with seismological observations (see sec. 1-2).

The phase diagram of $(Mg,Fe)_2SiO_4$, which is a representative mantle mineral, is shown in figure 1-6a. Minerals with this composition have the olivine crystal structure at low pressures, and they transform to wadsleyite (modified spinel), then to ringwoodite (spinel) at high pressures. At the pressure of about 24 GPa, they finally decompose into perovskite and magnesiowüstite. The gross picture of these phase transitions had been established almost completely by the mid-1980s.

While the phase diagram of iron, the main constituent of the core, has long been known under low-pressure conditions, it has not yet been completely known under the high-pressure and high-temperature field corresponding to the actual core conditions. On the basis of currently available theoretical and experimental grounds, iron with the hexagonal close-packed (hcp) structure seems to be the most likely material for the inner core (fig. 1-6b).

Once the phase diagram of a material is determined, the next step is to measure the density and elastic properties of each phase. This type of research has made rapid progress since the pioneering work by Birth (1952), and the measurements for core materials as well as mantle materials have been conducted. Particularly important in these studies are the development of new techniques of the generation of high pressures and temperatures and the development of techniques of measurements of properties under extreme conditions (from small samples, usually less than 1 mm³). Earth scientists have taken the lead in these areas, and these technologies often contribute to materials science and engineering, such as the development of new hard materials. The results of experimental studies have demonstrated that the depth variations of density and elastic properties

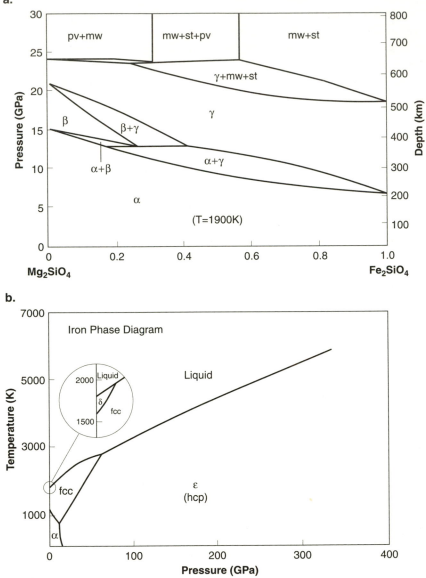

a.

b.

Fig. 1-6. Representative phase diagrams for Earth materials. (a) A phase diagram of $(Mg,Fe)_2SiO_4$ (Ringwood 1991) and (b) a phase diagram of Fe (Anderson 2002). A phase diagram for inner-core pressures and temperatures is still incomplete.

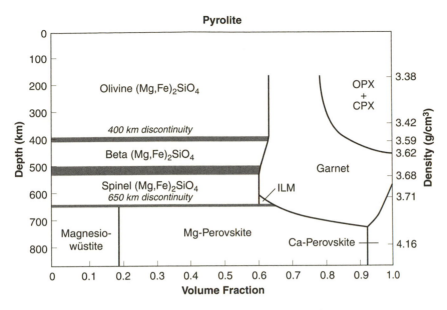

Fig. 1-7. A model for Earth constituent materials (after Ringwood 1991).

within Earth can be explained mostly as the effect of compression and abrupt phase transformations at some particular depths. The depth variation of major constituents for the mantle and the core, as estimated from these experimental results and seismological observations, is shown in figure 1-7.

There are two important points regarding phase transformations. The first is that the pressure at which a phase transformation takes place varies with temperature (temperature dependency), and the second is the possibility of non-equilibrium phase transformation.

a. Temperature Dependence of Phase Transformation

Many of phase transformations that take place within Earth are caused by the increase in pressure, so pressure is the most important variable. The effect of temperature, however, cannot be ignored. Thus, the pressure at which a given phase transformation occurs depends on the temperature. Therefore, when one plots the stable phases on a pressure-temperature plane, the boundary between the stability fields of two phases has a slope. This slope, $\left(\frac{dP}{dT}\right)_{eq}$, is called the Clapeyron slope, after the nineteenth-century French physicist who first constructed such a diagram (box 1-3).

Box 1-3. Phase Transformations and the Clapeyron Slope

A given substance, say H_2O, assumes various structures (phases) dependent on thermodynamic conditions (pressure, temperature, etc.). At room pressure, H_2O will be liquid water if the temperature is between 273 K and 373 K; below 273 K, it will be solid ice (ice I). This transition between liquid water and solid ice (ice I) will occur at different temperatures at different pressures. Similarly, carbon will assume graphite structure at room temperature and pressure, whereas it transforms to a diamond structure at higher pressures. The stability of a material under different conditions is determined by the Gibbs free energy, G,

$$G = U + PV - TS \tag{B1-3-1}$$

where U is internal energy, V is volume, and S is entropy (P is pressure and T is temperature). A given material will assume a structure with the lowest Gibbs free energy. Therefore a material assumes a structure with a smaller volume (i.e., higher density) at higher pressures, and a structure with higher entropy at higher temperatures. French physicist Clapeyron invented a way to show the stability of various phases, called a phase diagram. If we choose temperature (T) and pressure (P) as independent variables, then a P-T plane will be divided into several regions, each region corresponding to a stability field of a given phase. The slope of the boundary between the stability fields of two phases is called the Clapeyron slope.

Consider two adjacent points across a phase boundary corresponding to a temperature and pressure of (T, P). Because the two phases are in equilibrium at this boundary (T, P), the Gibbs free energy of the two phases (1 and 2) must be equal. Thus,

$$U_1 + PV_1 - TS_1 = U_2 + PV_2 - TS_2. \tag{B1-3-2}$$

At a nearby point ($T + dT$, $P + dP$), we have a similar relationship:

$$\begin{aligned} U_1 + (P + dP)V_1 - (T + dT)S_1 = \\ U_2 + (P + dP)V_2 - (T + dT)S_2. \end{aligned} \tag{B1-3-3}$$

From equations (B1-3-2) and (B1-3-3), we have

$$\left(\frac{dP}{dT} \right)_{eq} = \frac{S_1 - S_2}{V_1 - V_2} \tag{B1-2-4}$$

where *eq* is used to clearly indicate that this slope is for equilibrium between the two phases. This is the *Clapeyron slope*. The Clapeyron slope shows how temperature affects the pressure at which a given phase transformation occurs. If phase 1 is a high-pressure phase and the phase 2 is a low-pressure phase, then $V_1 < V_2$. But the relative magnitudes of entropy of the two phases can vary from one case to another. Usually, a high-pressure phase has a stiffer structure and has lower entropy than a low-pressure phase ($S_1 < S_2$). In such a case, the Clapeyron slope has a positive value. However, a phase transformation to a high-pressure phase can reduce the strength of the chemical bond in some unusual cases. In these cases, the Clapeyron slope has a negative value.

The Clapeyron slope is related to the change in volume and entropy associated with the phase transformation—namely,

$$\left(\frac{dP}{dT}\right)_{eq} = \frac{S_1 - S_2}{V_1 - V_2}, \tag{1-3}$$

where S_1 is the molar entropy of phase 1 (the high-pressure phase), S_2 is the molar entropy of phase 2 (the low-pressure phase), V_1 is the molar volume of phase 1 (the high-pressure phase), V_2 is the molar volume of phase 2 (the low-pressure phase), respectively. While $V_2 - V_1 > 0$ because the high-pressure phase always has a smaller volume than the low-pressure phase, the difference in entropy depends on the nature of each phase transition. Entropy is a key concept in thermodynamics that represents the degree of *disorder* of a system. The lattice vibration of atoms has the greatest influence on the entropy of solids. A higher frequency of lattice vibration corresponds to lower entropy. The high-pressure phase usually has stronger chemical bonding, which results in a higher frequency of atomic vibration, so it has less entropy than the low-pressure phase ($S_2 - S_1 > 0$). Thus $(dP/dT)_{eq} > 0$. When the coordination number (the number of atoms adjacent to a particular atom) is changed significantly by a phase transformation, however, the Clapeyron slope can be negative—for example, the phase transformation from ringwoodite to perovskite and magnesiowüstite (fig. 1-8). In this case, a silicon atom is surrounded by four oxygen atoms in ringwoodite but by six oxygen atoms in perovskite. Thus, the Si-O bonding in perovskite is relatively weak, and its entropy becomes greater, resulting in negative ΔS,—that is, $(dP/dT)_{eq} < 0$. Re-

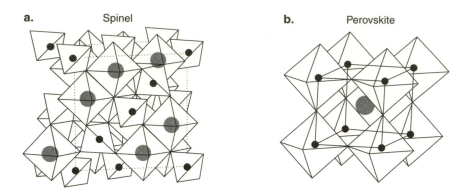

Fig. 1-8. The crystal structures of (a) spinel and (b) perovskite. Small dark circles represent Si, and large hatched circles are Mg (or Fe). (The perovskite structure shown here is the ideal cubic structure.) Note that Si atoms in spinel are located in the tetrahedra surrounded by four oxygen atoms, whereas Si atoms in perovskite are located in the octahedra surrounded by six oxygen atoms.

markably, this negative Clapeyron slope associated with the transformation to perovskite was predicted by the chemist Alexandra Navrotsky (then at University of Arizona, now at the University of California at Davis) (1980) and later experimentally demonstrated by Ito and Yamada (1982). Because of this temperature dependence of phase-transformation pressure, the depth of a phase transformation can vary from place to place if there are regions with anomalous temperatures.

This temperature dependence of phase transformation leads to an important effect. Consider, for example, the depth of the 660-km boundary. At around this depth, ringwoodite transforms to perovskite and magnesiowüstite. Because this phase transformation has a negative Clapeyron slope, it takes place at depths deeper than 660 km in regions with low temperature anomalies. This means that a less-dense phase (ringwoodite) extends into greater depths there than in other normal places, and this results in additional buoyancy, which tends to prevent the colder materials from sinking. The topography of this phase boundary caused by the lateral variation in temperature and therefore acts as a resisting force for convection (chap. 4). Although we usually use terms like the 410-km boundary and the 660-km boundary, the depths of these boundaries are actually different at different places. The fine structure of these phase boundaries has become observable in the 1990s (Shearer and Masters 1992).

b. Non-equilibrium Phase Transformation

In most of Earth's deep interior, the temperature is so high that phase transformations take place near chemical equilibrium. Inside a cold subducting plate, however, the temperature is much lower than the average mantle temperature. Phase transformations take place very slowly in these cold regions, so non-equilibrium phase transformation is possible. This type of non-equilibrium phase transition is frequently seen in our daily life. The best example is the diamond. Diamonds, which are one form of carbon, are unstable at room temperature and pressure. Though they should transform to graphite, this does not happen because the rate of phase transition at room temperature is extremely slow. A similar case can be found for silicates. For example, the depth of the olivine-wadsleyite transformation, which has a positive Clapeyron slope, should be shallower than average (410 km) for regions with low-temperature anomalies. If the temperature is very low, however, phase transformation may not take place at equilibrium, and olivine may be brought much deeper, until the phase transformation finally occurs. Although this possibility was already pointed out in a pioneering paper by Sung and Burns (1976), we were not able to discuss it quantitatively until Dave Rubie, now at Bayreuth in Germany, and his colleagues conducted detailed experimental research using a synchrotron radiation facility (box 1-4) in Japan (Rubie et al. 1990). Based on these results, some people consider that non-equilibrium transformation may occur in a subducting slab and that the non-equilibrium phase transformation is a cause for deep earthquakes (see chap. 5).

C. The Layered Structure of Earth: The Second-Order Approximation—Chemical Structure of the Mantle and the Core

In the previous, first-order approximation, the crust, the mantle, and the core are assumed to be chemically homogeneous. This assumption does not hold at a higher-order approximation. A prominent example is the core. The density of the outer core is lower than that of pure iron (or iron-nickel alloy) by about 10%, suggesting that it contains impurities in large quantities. On the other hand, the density of the inner core is close to that of pure iron (or iron-nickel alloy). Thus, the inner-outer core boundary is thought to be due not only to the change in phase (liquid and solid), but also to the change in composition. A plausible model to explain these phase and compositional differences is that the inner core has grown from materials in the outer core as a result of cooling. Because solubility of im-

Box 1-4. Synchrotron Radiation and New Mineral Physics Studies

Synchrotron is one of the particle accelerators designed for high-energy physics research (fig. B1-4-1). Charged particles (such as electrons) are accelerated by a huge ring of magnets and when they collide, strong X rays are emitted. The strength of these X rays is many orders of magnitude higher than the X rays generated by a conventional device, making them highly useful for research in many areas, including high-pressure mineral physics and medical science. In high-pressure mineral physics research, small samples

Fig. B1-4-1. The synchrotron facility at Argonne National Laboratory. Along a large circular ring (~ 150 m radius), charged particles are accelerated and upon collisions they emit high-energy X-rays. These X-rays can be used to investigate the properties of materials under high pressures.

surrounded by other materials that often absorb X rays must be investigated. Therefore, strong X rays are essential for these studies. A measurement that would take several days with a conventional X-ray source can be made within a few seconds with a synchrotron radiation facility. As a result, a number of experimental studies which were impossible or difficult are now possible with this powerful new facility. Many physical properties can be investigated through the use of the synchrotron facility, including phase relationships, elastic constants, densities (equation of state), viscosity of melt, plastic (rheological) properties, and the kinetics of phase transformations. These facilities are located in national laboratories such as Brookhaven National Laboratory and the Argonne National Laboratory in the United States, KEK (Ko-Energy-Ken) and Spring8 in Japan, and Grenoble in France.

purities is lower for solids than for liquids, impurities are accumulated in the outer core. According to this model, the growth of the inner core releases latent heat by solidification as well as gravitational potential energy by the removal of impurities, which provides significant energy for convection in the outer core (chap. 6).

How about the mantle? Is the mantle chemically homogeneous? As I explained before, most of the mantle structure, especially the structure of the transition zone, can be attributed to a first order, to the phase transformations of constituent minerals. This does not mean, however, that all of the mantle structure can be explained by changes in the physical properties of an isochemical material. Similarly to the core, chemical stratification is possible in the mantle due to the melting process. Differences between continental mantle and oceanic mantle are generally considered to be due to chemical heterogeneity. By combining seismic and gravity observations, Tom Jordan, then at Scripps Oceanographic Institution in California, concluded that the continental upper mantle is colder than the oceanic upper mantle, but that the density of the continental upper mantle is about the same as that of the oceanic upper mantle (1975). He suggested that this is due to the difference in the chemical composition between the continental upper mantle and the oceanic upper mantle: because the continental mantle, compared to the oceanic mantle, experienced more extensive partial melting, it is relatively depleted in dense minerals like garnet (see chap. 2).

Whether there is compositional heterogeneity in the deep mantle or not is an important problem related to the chemical evolution of the mantle, but it is still highly controversial. From geochemical observations, we know that there are at least two or more regions with different chemical compositions, and that these regions have not been mixed very much for more than one billion years (e.g., Hofmann 1997) (chap. 4). One hypothesis is that these different regions are the upper and lower mantle. In this case, a part of the difference in physical properties between the upper and lower mantle is due to compositional difference. In fact, some scientists have suggested that the density difference between the upper and lower mantle is due to the difference in the concentration of iron and other elements (Jeanloz and Knittle 1989). Others have proposed that the difference lies in silica concentration (e.g., Stixrude et al. 1992). Some consider that 410 km, not 660 km, is the chemical boundary. Don Anderson (1989) and Carl Agee (1993), argue that seismological data are consistent with the idea that the region between 410 and 660 km has a garnet-rich chemical composition. In principle, these hypotheses can be tested if we conduct accurate measurements of physical properties and compare them to seismological data. The problem is that differences in physical properties among different models are small, and the differences among the various models themselves are within the uncertainties in experimental and seismological observations. Furthermore, these properties (e.g., density) depend on temperature, which itself has some uncertainty, so it is difficult to arrive at a definitive conclusion.

In this type of argument, the main issue has usually been whether the upper and lower mantle have the same chemical composition. Recently, Rob van der Hilst and H. Karason (1999), at the Massachusetts Institute of Technology (MIT), suggested that a chemical boundary may exist at ~ 1,600 km depth, not at the upper-lower mantle boundary (660 km). Their argument is based on the anticorrelation, in their seismic tomography, between the bulk modulus and the shear modulus observed deeper than ~ 1,600 km, which cannot be explained as the effect of temperature alone (I will explain this in more detail in chap. 3).

Though small in quantity, water distribution in the mantle is also heterogeneous. The distribution of water can be estimated by chemical analyses of igneous rocks. Water is abundant at island arcs such as Japan and less abundant at mid-ocean ridges. Water is also redistributed by the partial melting of rocks; a large volume of water can be dissolved into melt (molten rocks), whereas very little water can be dissolved in solid minerals. Thus, partial melting results in the drying out of minerals because water in minerals is absorbed into melt. Water can greatly modify the viscosity of rocks, so the distribution of water is important for geodynamics. It is also considered to have a large effect on the layering of the lithosphere and the asthenosphere. This issue will be discussed in more detail in chap. 2.

Though the origin of chemical heterogeneity of the mantle is yet uncertain, its presence is strongly supported by geochemical observations. The question is how chemically different materials are distributed spatially and temporally. We have just started to work on this problem on the basis of observation. It is a fundamental problem related to the evolution of Earth and its dynamics, and significant progress in the future is expected. Chapters 3 and 4 of this book deal with this problem in some detail.

1-2. THE THERMAL STRUCTURE OF EARTH'S INTERIOR

In the previous discussion on the composition of Earth's interior, the issue of temperature was not so important because density and elastic properties vary little by temperature. However, some properties such as viscosity can strongly depend on temperature. Because viscosity is one measure of the mobility of materials, temperature is expected to have a great effect on flow patterns in Earth's interior. Conversely, the temperature distribution also depends on the pattern of convection and the distribution of the heat source. If we can estimate the temperature distribution, the results can provide useful constraints on geodynamics such as the convective pattern.

The temperature gradient near the surface can be measured by measuring the temperatures in a deep well. These measurements show that the temperature gradients are about 10–50 K/km, although they can differ from one region to another. Heat transfer near the surface occurs mostly by thermal conduction, so we may assume a roughly constant temperature gradient to some depth. If the temperature in the deep portion is estimated this way, the temperature at the depth of 200 km is estimated to be 2,000–10,000 K. Since these temperatures exceed the melting temperatures of rocks, rocks must be largely molten at these greater depths. This inference is, however, inconsistent with the observation that shear waves propagate through the mantle. We must conclude that our assumption of (nearly) constant temperature gradient is wrong; a temperature gradient in the deep interior should be much smaller than that near the surface.

This conclusion is also supported by the following argument. As ex-

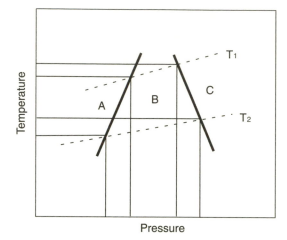

Fig. 1-9. The Clapeyron slope and the depth and temperature of phase transformation. Phase transformation A-B has a positive Clapeyron slope, and B-C has a negative one. By observing phase transformation depth by seismology, temperature at that depth can be estimated. T_1 and T_2 are hypothetical geotherms.

plained before, the pressure at which a phase-transformation pressure occurs depends on temperature. Seismic observations tell us that there are seismic discontinuities at the depths of 410 km and 660 km. If these discontinuities are caused by phase transformations, these depths can be used to estimate the temperatures at the discontinuities (fig. 1-9). By this approach, Eiji Ito and Tomoo Katsura, at Misasa in Japan (1989) estimated the temperature of the mantle transition zone and showed that the temperature gradient there ($\approx$ 0.5 K/km) is much smaller than the surface value. This variation in temperature gradient can also be supported by an argument using the Bullen parameter, shown in equation (1-2).

Though we do not know for certain the temperature below 660 km, we can at least place several important constraints. For example, at the depth of 5,150 km (the inner-outer core boundary) the temperature must be the melting temperature of iron. So we can infer the temperature there by measuring the melting temperature of iron. This type of measurement is, however, very difficult, and the best estimate is ~ 5,000 K with a large margin of error, of 500 K or more. This estimate is still useful, and we can draw an important conclusion on the temperature distribution and heat-transfer mechanism in Earth's interior. In the outer core, where the magnetic field is believed to be generated by a dynamo, we can assume vigorous convection; its temperature gradient is close to the adiabatic thermal gradient (around 0.6–0.8 K/km). This gradient gives 3,600 ± 500K for the uppermost part of the core (i.e., the core-mantle boundary). This is significantly higher than 2,500 K, which is calculated from the estimated temperature at 660 km (1,900 K), assuming the adiabatic thermal gradient. From this discrepancy, we can conclude that there must be a high temperature gradient at the lowermost mantle (called the D″ layer).

This is the current standard model for temperature distribution in the mantle (fig. 1-8a). In this standard model, thermal boundary layers exist only at the uppermost and lowermost parts of the mantle. This type of temperature distribution is expected when not only the mantle but also the core act as heat sources and when mantle convection is of whole-mantle scale. In this case, there is a large temperature gradient at the core-mantle boundary, so plumes are likely to form there.

To understand the thermal structure in terms of the dynamics of Earth's interior, we need to know some fundamentals of heat transfer. Thermal conduction and thermal convection are two important heat-transfer mechanisms in Earth's interior (heat transfer by radiation may also be important in the very hot regions). Heat transfer by convection occurs when a material is carried to a position where the temperature is different from that of a material being transported. Vertical convective heat transfer is not expected in the lithosphere (a shallow cold region), so vertical heat transfer occurs mostly by conduction within the lithosphere. Conduction is not an efficient mechanism of heat transfer, so its thermal gradient is large. At greater depths, viscosity becomes low due to the higher temperature, and (vertical) heat transfer by convection becomes more efficient. The temperature gradient for convection is much smaller than that for conduction and is close to the adiabatic thermal gradient (0.3–0.4 K/km). The basic physics for this change in heat-transfer mechanism, from thermal conduction to convection, can be understood as follows: Convection takes place by temperature difference in either a horizontal or a vertical direction. Let us focus here on the vertical temperature difference because we are concerned with heat transfer from the deep mantle to the surface. Consider a layer of viscous fluid heated from below. The deeper part of this layer becomes lighter by heating and starts to rise. As it rises, it loses heat and hence buoyancy through conduction of heat to the surrounding colder fluid. Continuous vertical fluid motion—that is, thermal convection—is possible only when the time-scale for thermal conduction is larger than that for vertical advection, so that fluid temperature does not change so much during its vertical motion. Lord Rayleigh, a British physicist, analyzed this problem in the early-twentieth century and found a condition under which thermal convection occurs. The condition for thermal convection is given by a nondimensional number, the Rayleigh number, defined by

$$Ra \equiv \frac{\rho g h^3 \alpha \Delta T}{\eta \kappa},$$ (1-4)

where α is thermal expansion, ρ is density, ΔT is the temperature difference between top and bottom boundaries, g is acceleration due to gravity, h is layer thickness, k is thermal diffusivity, and η is viscosity. The Rayleigh number denotes the ratio of the thermal diffusion time-scale to the vertical advection time-scale. Convection takes place when the Rayleigh number exceeds a certain value (approximately 1,000). Viscosity is the most uncertain quantity when evaluating the Rayleigh number for the mantle, but it is generally thought to be about 10^{20}–10^{22} Pa·s on the basis of various kinds of geophysical inference, as I will discuss later. With this range of viscosity, the Rayleigh number is 10^5–10^7 for the mantle. Therefore we conclude that the mantle is vigorously convecting. The thermal structure and flow pattern of an intensely convecting fluid layer can be well explained by the *boundary layer theory*. At high Rayleigh numbers, the buoyancy forces driving convection are concentrated in the top and bottom boundary layers, and fluids slowly move vertically in between. Within the boundary layers, fluid motion is almost horizontal, and a vertical temperature gradient there is controlled by thermal conduction. The boundary layers have a large temperature gradient, which is determined by the layer thickness, thermal conductivity, and heat flux through the layer. In most of the fluid layer, on the other hand, vertical fluid motion is significant and heat is transported by advection, and hence the temperature gradient is adiabatic (0.3–0.4 K/km). Thus, typical temperature distribution in a fluid layer becomes the one shown in figure 1-10a. Note that the top boundary layer is well known as the lithospheric plate, but the bottom boundary layer at the base of the mantle is still elusive. The temperature gradient at the bottom layer is determined by the heat flux from the core, which is quite uncertain.

There are a few uncertainties about this standard model. The first one is whether a boundary layer exists between the upper and lower mantle. As I will explain later in detail, there are a growing number of observations that cast doubt about a simple model of whole-mantle convection. If convection is separated into the upper and lower mantle, there should be a thermal boundary layer between them, and the thermal gradient there should be larger than the adiabatic value (fig. 1-10c). The second uncer-

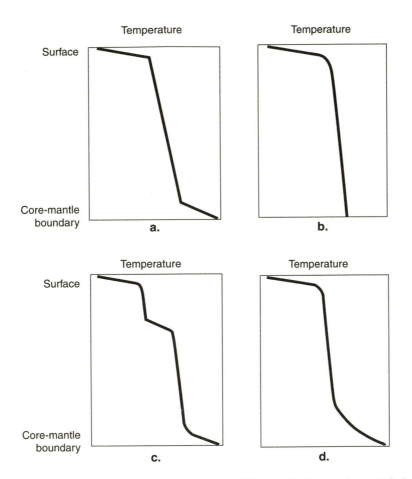

Fig. 1-10. Temperature distribution in the mantle and the mode of convection. (a) Whole-mantle convection: the case of a large heat flux from the core. Thermal boundary layers are formed at both top and bottom boundaries. (b) Whole-mantle convection: insignificant heat flux from the core. The thermal boundary layer is formed only at the top boundary. (c) Layered convection: large core heat flux is assumed. The thermal boundary is formed also at the boundary of two layers. (d) Whole-mantle convection with less vigorous convection at the deep mantle. Thermal gradient at the weak convection region is higher than the adiabatic thermal gradient.

tainty is related to the temperature gradient in the lower mantle. Recent studies seem to indicate that the thermal gradient in a significant portion of the lower mantle is larger than the adiabatic value. On the basis of their seismic tomography, van der Hilst and Karason (1999) suggested that there is a chemically distinct layer below ~ 1,600 km, which does not mix with the shallower region. In this case, because there is no vertical fluid motion below ~ 1,600 km, the temperature gradient there should be larger than the adiabatic value (fig. 1-10b). This possibility of a superadiabatic thermal gradient in the lower mantle is also supported by the radial distribution of elastic constants and density as estimated by seismological studies and by the radial variation of viscosity (which I will explain later).

In the above arguments, we have considered only the radial structure of temperature. Of course, temperature in a convecting fluid varies also in a horizontal direction. Temperature beneath mid-ocean ridges where hot-mantle materials rise is much higher than that beneath old cratonic continents. Similarly, there seems to be a large lateral thermal gradient at the core-mantle boundary.

1-3. RHEOLOGICAL STRUCTURE: SEISMIC WAVE ATTENUATION AND VISCOSITY

In addition to elastic properties, anelastic properties and viscosity are also important for geodynamics. In this section, I will summarize the distribution of seismic wave attenuation and viscosity. Both seismic wave attenuation and viscous deformation occur through non-elastic deformation and they are often collectively called *rheological properties* (*rheo* means "flow" in Greek) (box 1-5). A seismic wave is an elastic wave, but its amplitude decreases as it propagates because elastic energy is converted into heat. This conversion occurs because Earth materials are not a perfect elastic body and have some viscous character. This behavior of material is called *anelasticity*. Briefly, anelasticity is a type of mechanical behavior of a material that is between elastic and viscous behavior. If Earth were a perfect elastic body, it would continue to oscillate forever after an earthquake, and this would give us a lot of trouble. Fortunately, seismic oscillation eventually stops because Earth is not a perfect elastic body.

Because viscous behavior in solids occurs only at high temperatures, the anelastic effect in solid rocks is particularly prominent at high temperatures and low frequencies. Therefore, the effects of anelasticity are negligible for the high-frequency (100–1,000 MHz) elastic waves usually used in laboratory experiments. However, seismic waves have low frequencies (1–0.001 Hz), so this anelastic effect cannot be neglected. The anelastic effect is expressed by Q, where Q^{-1} denotes the fraction of elastic vibration energy dissipated as heat. A smaller Q means low attenuation. In the mantle, Q is on the order of 50–500. That is, about 1–10% of energy is lost during one cycle of vibration. Q has a close relation to viscosity, and

Box 1-5. The Atomistic Basis of Elastic and Plastic (Viscous) Deformation

In solid Earth geophysics, we often deal with the deformation of materials by an external force. When a small force is applied at a relatively low temperature and for a short time, deformation occurs instantaneously. As soon as the force is removed, the material goes back to its original undeformed state. This style of deformation is called *elastic* deformation. Deformation associated with the propagation of seismic waves is nearly elastic. In contrast, deformation associated with long-term gravitational forces (caused by buoyancy forces) in the hot mantle occurs *viscously*. Deformation in this case occurs slowly, and even after the removal of force, permanent deformation (strain) remains.

The difference between the two modes of deformation can be understood at the atomic scale. Recall that each solid is made of a periodic array of atoms. Each atom sits at each position where the interaction energy between the atoms is at the minimum. When a small force is applied, atoms move slightly out of their stable (equilibrium) position, which creates a restoring force. When the displacement is small, the restoring force is proportional to the distance of movement and to the *curvature* of the interatomic potential. When the force is removed, this restoring force brings an atom back to its original position. The curvature of the interatomic potential changes slightly with temperature (fig. B1-5-1). At high temperatures, the curvature becomes small and the restoring force is weak (effects caused by the change in the curvature of the interatomic potential are referred to as *anharmonic effects*). Therefore, the elastic constants decrease slightly with temperature, and the dependence is usually nearly linear. The degree of atomic motion can be higher when a large force is applied or a force is applied for a long time. In these cases, atoms can move over the potential maximum to neighboring positions. Once atoms move to neighboring positions, then even after the removal of the force, they will not automatically move back to the original positions: the deformation is permanent (plastic deformation). Consider a case where a small force is applied for a long time at a high temperature. At high temperatures, all atoms vibrate—that is, they do not remain at their equilibrium positions—but by statistical fluctuation, their positions move around their equi-

(*continued*)

librium positions. Consequently, there is a finite probability that an atom can go over the potential hill to the neighboring position. According to the analysis by Austrian physicist Ludwig Boltzmann, this probability is dependent upon the amplitude of atomic vibration; hence, temperature as $\sim exp[-H^*/RT]$, where T is temperature, R is the gas constant, and H^* is the activation enthalpy (the height of the potential hill). Thus, the rate at which plastic deformation occurs is proportional to $\sim exp[-H^*/RT]$ and is highly sensitive to temperature. For a typical H^* of ~ 500 kJ/mol, a 100-degree increase in temperature (at $T = 1,600$ K) causes an increase of the rate of deformation by a factor of ~ 10.

Actually, the atomic jump that causes plastic deformation is facilitated by the presence of *lattice defects* such as crystal dislocations and vacant lattice sites (vacancies). Some details of mechanisms of plastic deformation are discussed in chapter 2.

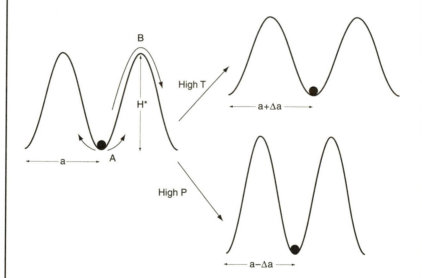

Fig. B1-5-1. A change in interatomic potential with temperature (T) and pressure (P). The curvature of the potential determines the magnitude of the restoring force for small displacement (A) and, hence, elastic constants. With higher temperatures (pressures), the mean atomic spacing, a, increases (decreases) due to thermal expansion (compression), and the curvature of the potential becomes smaller (larger). With a certain probability, atoms can also move into the next stable position (B). This causes permanent (plastic) deformation. The probability of this atomic jump is determined by the height of potential barrier (H^*) and the magnitude of atomic vibration (i.e., temperature), as well as stress magnitude.

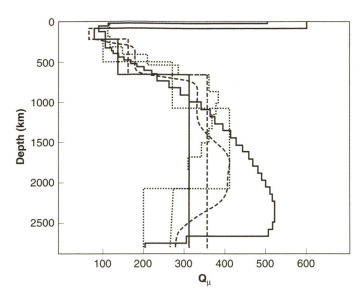

Fig. 1-11. Depth variation of Q_μ (seismic attenuation parameter for shear waves) (after Romanowicz and Durek 2000).

it depends on temperature, pressure, and frequency. Because Q is related to viscosity, it also depends on impurities such as water.

Seismic wave attenuation can be measured by the amplitudes of waves and by the width of peaks in the spectrum of Earth's free oscillation. Figure 1-11 shows a model in which Q is assumed, for simplicity, to be a function of depth only. The Q is large and attenuation is low near the surface. Attenuation is high at depths of 100–300 km. At greater depths in the mantle, attenuation becomes lower again.

Thus, the viscous character of Earth is already seen at the frequency range for seismic waves. At much lower frequencies (or a longer time-scale), the viscous character becomes much more pronounced. By analyzing such long time-scale phenomena, one can estimate the viscosity variation within Earth. About 10,000 years ago, thick ice sheets covered regions in the northern hemisphere. These ice sheets quicky melted about 6,000 years ago, and the regions that had been covered with ice started to rise slowly. By theoretically analyzing this crustal uplift, mantle viscosity can be calculated. This type of research was first conducted in 1935 by Norman Haskell, at MIT, and others. In the latter decades of the twentieth century Dick Peltier, Jerry Mitrovica, and Alessandro Forte (now at the University of Western Ontario), at the University of

Box 1-6. Dimensional Analysis for Relaxation Time for Post-Glacial Rebound

The postglacial rebound (slow vertical crustal movement) after the melting of glacier occurs by gravity force and is controlled by the viscosity of Earth materials. Therefore, its time-scale (relaxation time) (τ) is controlled by gravity (g), density (ρ) and viscosity (η), as well as the size (λ) of the regions of melting (because the gravity force depends on the size). Thus, one can write,

$$\tau \propto g^{\alpha} \eta^{\beta} \lambda^{\gamma} \rho^{\delta}, \tag{B1-6-1}$$

where α, β, γ, and δ are constants. The quantities of both sides of this equation must have the same unit. Therefore,

$$[s] = [m \cdot s^{-2}]^{\alpha} [kg \cdot m^{-1} \cdot s^{-1}]^{\beta} [m]^{\gamma} [kg \cdot m^{-3}]^{\delta}, \tag{B1-6-2}$$

where [] stand for the unit, and hence

$$\alpha - \beta + \gamma - 3\delta = 0,$$
$$-2\alpha - \beta = 1, \text{ and} \tag{B1-6-3}$$
$$\beta + \delta = 0.$$

Solving these equations with a condition $\alpha = \delta$ (because the gravitational force is always in the form ρg), one gets $\alpha = \delta = -1$ $\beta = 1$ and $\gamma = -1$. Thus,

$$\tau = (\rho g \eta / \lambda) \cdot F, \tag{B1-6-4}$$

where F is a nondimensional constant.

Toronto in Canada; Masao Nakada (now at Kyushu University, Japan); and Kurt Lambeck, at ANU, made important contributions to this approach.

Suppose that a surface load is changed at some time. Because hydrostatic balance is then lost, a medium starts to flow. In the case where a load is suddenly removed, a medium flows into a place where a compressive load had formerly been applied, leading to crustal uplift around the region. The time-scale of this uplift depends on the viscosity of the medium. Thus, by measuring the uplift as a function of time, viscosity can be estimated. A dimensional analysis shows that this time-scale τ depends on viscosity as (box 1-6),

$$\tau = \frac{\eta}{\rho g \lambda} F, \qquad (1\text{-}5)$$

where λ is the horizontal length scale of the load, ρ is density, g is acceleration due to gravity, and F is a nondimensional number. F is a function of λ, when viscosity is depth-dependent (e.g., $F \approx [\lambda/H]^3$) in the presence of a low-viscosity layer whose thickness is H). If viscosity is assumed to be independent of depth, $F = 4\pi$, and the average mantle viscosity is estimated to be 3×10^{21} Pa·s using the observed values of τ and λ. If viscosity can vary as a function of depth, however, the expression for F becomes complicated, and it is difficult to determine with precision the depth variation of viscosity. In addition, because this approach is based on deformation due to a surface load, it is difficult to determine viscosity in the deep mantle. Consequently, the viscosity in the deep regions of Earth's mantle had been poorly determined and controversial. In the 1970s until the late 1980s, a majority of scientists argued, based on the analysis of postglacial rebound, that the viscosity of the mantle is nearly independent of depth (Cathles 1975; Peltier 1989), until Jerry Mitrovica and Dick Peltier showed that the viscosity of the mantle deeper than ~ 1,200 km cannot be estimated by this method (1991). The work by Masao Nakada and Kurt Lambeck was an exception. Through a thorough analysis of the postglacial rebound incorporating the important effects of coastline geometry, which had been ignored by most of the previous works, Nakada and Lambeck (1989) showed a significant increase in viscosity with depth.

A completely different method for determining the depth variation of viscosity uses gravity observation to investigate how density anomalies in the mantle are maintained by the viscous flow. In the dynamic Earth's interior, mass anomaly is maintained not by static elastic deformation but by dynamic viscous flow. Because of this, even when there is a dense material beneath the surface, we do not always observe a positive gravity anomaly (stronger gravitational field) at the surface. Viscous flow causes surface deformation, and the resultant *dynamic topography* (topography caused by viscous flow) has a large effect on gravity. Consequently, when mass anomalies are maintained by viscous flow, surface gravity anomalies are sensitive to the depth variation of viscosity as well as the mass distribution within Earth. By comparing observed gravity anomalies (more precisely, anomalies in the equi-potential surface of gravity field, or *geoid*) with those calculated from theoretical models of flow corresponding to a particular viscosity-depth profile, we can estimate the depth profile of viscosity. This method was developed in the mid-1980s by Brad Hager and Mark Richards, then at Caltech (now at MIT and the University of California at Berkeley, respectively) (Hager 1984; Richards and Hager 1984). This important pioneering work has been followed by a number of later studies. Their work gave the first physically sensible explanation for the observed large positive geoid anomalies in the western Pacific. In that region, cold (heavy) materials are sinking into the mantle beneath ocean trenches. Therefore, it might appear obvious that positive geoid anomalies occur there. However, Hager showed that if the mantle viscosity is independent of depth, a popular idea at that time, then one should expect negative geoid anomalies at the surface due to the strong effect of the depression of surface topography due to viscous flow (dynamic topography). The observed positive geoid anomalies in the western Pacific imply that the effects of dynamic topography are much less than expected from the fluid motion with a constant viscosity. In other words, the vertical fluid motion is more sluggish due, presumably, to the rapid increase in viscosity with depth. Using this approach, Hager (1984) was the first to clearly demonstrate that the viscosity of the lower mantle is much higher than that of the upper mantle. To use this method, density anomalies in the mantle must be known. Hager considered only the contribution of subducted oceanic slabs to density anomalies, and he calculated the density of subducted slabs from the estimated temperature difference. Later, as seismic tomography developed, it became popular to use seismic velocity anomalies to estimate density anomalies, and then to estimate viscosity variations. As I will explain in chapter 3, to estimate density anomalies from velocity anomaliesit is not a straightforward process; one of the limitations of this method lies in the uncertainty of estimated density anomalies. As a common issue for gravity-based inferences, there is also a non-uniqueness problem. More than one model can explain the same gravity anomalies.

Figure 1-12 shows the estimated viscosity variations by these methods. Although the estimation of viscosity is not quite as accurate as that of seismic velocity, we may summarize as follows:

1. The average mantle viscosity of the mantle is well constrained; it is around 3×10^{21} Pa·s.
2. At shallow mantle (100–300 km), there is a low viscosity layer, with

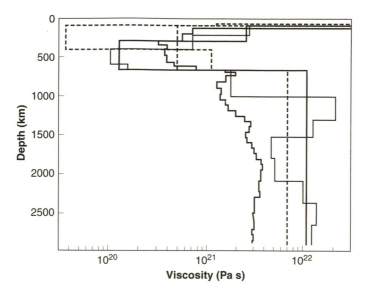

Fig. 1-12. Depth variation of viscosity (after Peltier 1998). All models consistently show that the upper mantle (asthenosphere) has low viscosity and the lower mantle has high viscosity. Transition zone viscosity as well as viscosity variation in the lower mantle are not very well determined.

viscosity of 10^{19}–10^{20} Pa·s, though it seems to vary from place to place.

3. The deep mantle (deeper than ~ 700 km) has a higher viscosity (10^{21}–10^{22} Pa·s) than the shallower mantle.

4. Some models suggest the existence of a low-viscosity layer in the mid-mantle. Without this layer, mantle convection seems to produce too large a surface deformation (dynamic topography). The details of this low-viscosity layer, such as its exact depth, are, however, not well resolved.

This type of viscosity structure roughly corresponds to the depth variation of seismic wave attenuation shown in figure 1-11. Both results show that the shallow mantle (~ 100–300 km) is weak (low viscosity) and the deep mantle is strong (high viscosity). Attenuation and viscosity are closely related because both of them are caused by similar microscopic processes (the motion of lattice defects) (see box 1-7 and chap. 2). With a reasonable range of parameters determined by laboratory studies, a fac-

Box 1-7. Seismic Wave Attenuation and Viscosity

Both seismic wave attenuation and the viscosity of rocks are caused by the slow motion of atoms over the potential hill (box 1-5) and are highly sensitive to temperature. Seismic wave attenuation is often measured by a Q-factor, which is defined by $Q^{-1} \equiv$ *[dissipated energy]/[stored energy]*, That is, a low Q means high attenuation. Laboratory experiments show that in most cases, Q depends on temperature and frequency, as

$$Q^{-1} \propto \omega^{-\alpha} exp[-\frac{\alpha H_Q^*}{RT}], \qquad \text{(B1-7-1)}$$

where ω is the frequency, R is the gas constant, T is the temperature, α is a constant ($\sim 0.2–0.3$) and H_Q^* is the activation enthalpy for attenuation. The rate of deformation (strain rate) in viscous deformation changes with temperature and stress (σ) as,

$$\dot{\varepsilon} \propto \sigma^n exp[-\frac{H_\eta^*}{RT}]. \qquad \text{(B1-7-2)}$$

Hence the viscosity, η ($= \sigma/\dot{\varepsilon}$), depends on temperature as

$$\eta \propto exp[\frac{H_\eta^*}{RT}] \quad or \quad \eta \propto exp[\frac{H_\eta^*}{nRT}] \qquad \text{(B1-7-3)}$$

for constant stress or a constant strain-rate, respectively, where H_η^* is the activation enthalpy for viscosity. From these equation, one has

$$Q \propto \eta^\beta, \qquad \text{(B1-7-4)}$$

with $\beta = \alpha H_Q^*/H_\eta^*$ for constant stress and $\beta = \alpha n H_Q^*/H_\eta^*$ for a constant strain rate. In most mantle materials, $H_Q^* \sim H_\eta^*$ and $\beta = 0.2–0.9$.

tor of 3 increase in Q corresponds to a $\sim 4 - \sim 200$-fold increase in viscosity. This roughly explains the correlation between the depth variation of attenuation shown in figure 1-11 and that of viscosity shown in figure 1-12.

Both attenuation and viscosity change with depth corresponding to the change in pressure and temperature (and crystal structure and possibly chemical composition) with depth. The pressure dependence of seismic wave attenuation and viscosity can be expressed in terms of the pressure dependence of activation enthalpy (box 1-5),

$$H^*_{Q,\eta} = E^*_{Q,\eta} + PV^*_{Q,\eta}, \tag{1-6}$$

where $E^*_{Q,\eta}$ is activation energy, $VE^*_{Q,\eta}$ is activation volume (for attenuation and viscosity respectively), and P is pressure. Activation volume represents the degree to which the potential barrier for atomic motion increases with pressure, and it is usually positive. Thus an increase in pressure decreases attenuation and increases viscosity. The overall depth variation of rheological properties shown in figures 1-11 and 1-12 can be explained by the effects of temperature and pressure. At shallower parts, temperature rapidly increases, so viscosity decreases with depth. Going even deeper, the pressure effect becomes more significant, so viscosity starts to increase with depth. In this argument, however, the effect of phase transformation is not considered. Phase transformations may significantly modify rheological properties in some cases.

In this type of global-scale estimate of viscosity, smaller-scale variations of viscosity are ignored. However, small-scale viscosity variations may be important in some situations. An example may be found in the dynamics of the subduction process. The subducted oceanic plate probably has high viscosity because it is much colder than the surrounding mantle. In the subducted slab, however, viscosity may change in a complex manner because of successive phase transformations and the variations in temperature. This issue is closely related to the fate of subducted slabs and whole-mantle circulation, and I will discuss it in detail in chapter 4. Another example is mid-ocean ridge dynamics. Partial melting redistributes water in the upper mantle, and this may cause a change in viscosity (chap. 2).

TWO • WATER, PARTIAL MELTING, AND THE LITHOSPHERE-ASTHENOSPHERE

2-1. LITHOSPHERE AND ASTHENOSPHERE: GEOPHYSICAL OBSERVATIONS

In the Earth model discussed in chapter 1, the upper mantle was treated as a single layer. However, if one looks at the structure of the upper mantle in more detail, it becomes obvious that the upper mantle can be divided into two mechanically distinct layers: the lithosphere (a strong layer; *lithos* means "rock"—i.e., strong but fragile—in Greek) and the asthenosphere (a weak layer; *asthenos* means "fluid" in Greek). These two layers are chemically nearly identical, but their mechanical properties are largely different. This mechanical layering is a key to the operation of plate tectonics on Earth.

We have known about such a mechanical layering for more than fifty years. Beno Gutenberg, a German seismologist who moved to the United States and, with Charles Richter, founded the Seismological Lab at Caltech, showed that seismic waves propagate more slowly through the depth range of ~ 50–150 km in the upper mantle than through the adjacent shallower and deeper regions (fig. 2-1). In his early paper on the seismic low-velocity zone, Gutenberg (1954) identified this layer with a layer of *partial melting*. Later, this layer was also shown to be associated with high seismic attenuation and high electric conductivity (e.g., Shankland, O'Connell, and Waff 1981). Its viscosity is also estimated to be lower than that of the upper and lower layers (fig. 1-11). Almost all standard textbooks on Earth science present this model of the partial-melting origin of the asthenosphere. However, based on recent studies of the properties of Earth materials, it now appears that this traditional view needs a fundamental revision, as I will explain in this chapter. In this new view, which is based on mineral physics studies, the asthenosphere is a layer where a significant partial melting does *not* occur. A key to this revision is the

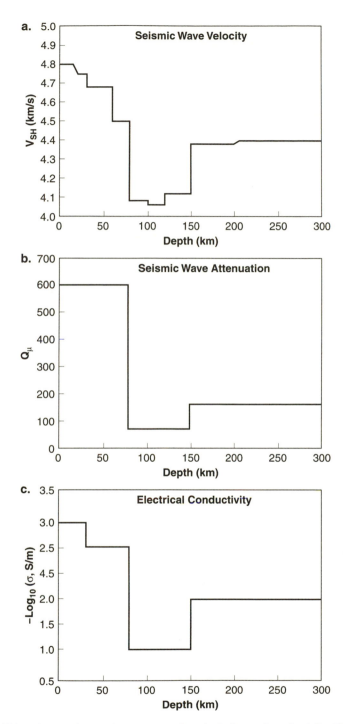

Fig. 2-1. Lithosphere-asthenosphere structure for a typical oceanic region (after Shankland et al. 1981).

demonstration, by laboratory studies, that a small amount of *water* has dramatic effects on the physical properties of minerals, whereas a small amount of partial melting does not have significant effects on mechanical properties, and that the distribution of water is modified by partial melting. In this chapter, I will first provide a historical review of studies on the origin of the low-velocity zone, and will discuss how water might be distributed in Earth's interior and might affect its properties.

The structures of the lithosphere and the asthenosphere vary significantly from place to place. While the well-defined asthenosphere is not recognized by the Jeffreys-Bullen's Earth model, which was based mostly on body wave data through continental regions, it is clearly seen in a model by Gutenberg, who mainly studied seismic wave propagation through oceanic regions. Later, in the 1970s, Toshikatsu Yoshii, at Earthquake Research Institute, University of Tokyo (1975), Don Forsyth, in the United States (1975), and others conducted detailed studies of the upper-mantle structure using surface waves, and it became apparent that the oceanic and continental regions have different structures of the lithosphere and the asthenosphere. These studies also showed that even in the same oceanic region, the structure varies for different ocean floor ages, and that old continents and young continents also have different structures. Surface wave studies show that the thickness of the seismic lithosphere, which is defined by high seismic velocities, is almost proportional to the square root of the ocean floor age (fig. 2-2).

Subtler features have also been identified through high-resolution studies using body waves, which have shorter wavelengths than surface waves. Jim Gaherty and Tom Jordan, then at Massachusetts Institute of Technology, for example, found seismic reflection at around 60–70 km depth in the oceanic upper mantle and discovered that this depth is similar between old oceanic mantle (western Pacific) and young oceanic mantle (in the Philippine sea) (Gaherty and Jordan 1995) (fig. 2-2). This suggests that the boundary must be sharp, having a transition width shorter than or comparable to the wavelength of seismic waves. The physical properties therefore must drastically change within a few kilometers. Nor does this boundary depend on the age of ocean floor. A similar, weakly age-dependent structure was also inferred by Regan and Anderson (1984), who used surface wave data incorporating anisotropy. The model by Forsyth, who also incorporated a simple version of the anisotropic model, shows a weaker age dependence of the structure of the oceanic lithosphere than does the model by Yoshii, who did not include the effects of anisotropy. In summary, the structure of the oceanic lithosphere (and the

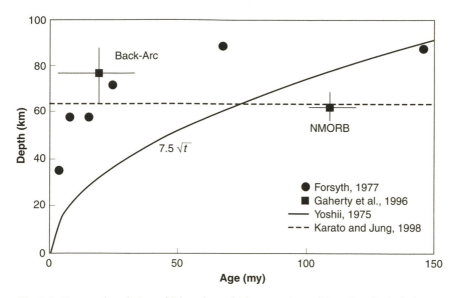

Fig. 2-2. Temporal evolution of lithosphere thickness estimated by seismological observations (after Karato and Jung 1998).

asthenosphere) depends on age, but the details of evolution depend on the technique used to infer the structure.

Structures of the upper mantle in the continental regions are quite different. A well-defined asthenosphere is not often observed. However, a thick high-velocity (seismic wave velocity) lithosphere is found in the continental regions, and the thickness of this lithosphere systematically varies with age: a very thick lithosphere in the Archean (older than ~ 2.7 billion years) regions and thinner ones in the Proterozoic (younger than ~ 2.7 billion years) regions.

In this chapter, I will discuss the origin of these structures in the upper mantle, and will show the importance of water in defining mechanical stratification.

2-2. THE MECHANISM OF THE FORMATION OF THE LITHOSPHERE-ASTHENOSPHERE STRUCTURE

2-2-1. The Evolution of the Oceanic Lithosphere: Classical Models

As soon as the theory of plate tectonics was proposed, quantitative models for the oceanic lithosphere were also formulated (e.g., McKenzie

1967). Important observations on which these models had been formulated included (1) ocean floor depths (bathymetry), (2) heat flow values, and (3) seismic wave velocities (fig. 2-1). Historically, the initial focus was on the data of heat flow, because heat is a key to understanding convection, but the interpretation of heat flow data turned out to be not so straightforward. Heat flow data near mid-ocean ridges are very scattered, and it took a long time to understand the cause of this scatter. In the late 1970s, hydrothermal circulation at mid-ocean ridges was discovered and its role in modifying heat flow was evaluated: we finally became able to quantitatively interpret heat flow data.

In contrast to the results of heat flow measurements, the data on sea floor depth (bathymetry) are simple and beautiful. The depth of ocean floor created up to about 80 million years can be remarkably well described by the following relationship (box 2-1):

$$d = d_o + a\sqrt{t}, \tag{2-1}$$

where d is the depth, d_o is the depth at age $(t) = 0$, t is the age of ocean floor (measured from its birth at the mid-ocean ridge), and a is a constant. The depths of the sea floor older than ~ 80 million years are, however, shallower than predicted by this equation. Sea floor depth can be interpreted as due to density anomalies using the principle of *isostasy*, so this relation reflects an increase in density due to the cooling of the lithosphere (box 2-1).

Most of the observations, particularly those on the bathymetry, can be explained nicely by a simple model of thermal evolution of the oceanic lithosphere by cooling (see box 2-1). I want to mention two subtleties regarding the evolution of the oceanic lithosphere. First are the boundary conditions at the bottom of the lithosphere. Physically the issue is, what defines the lithosphere? Dan McKenzie at Cambridge University, one of the founders of the theory of plate tectonics, proposed that the thickness of the oceanic lithosphere is constant (1967). An idea behind this assumption is that the thickness of the lithosphere is controlled by the melting temperature of mantle materials so that a constant temperature at the bottom of plate is maintained. Even with this constant thickness model, one can explain the age dependence of bathymetry and other quantities because the temperature (and density) in the lithosphere changes with age due to cooling. Toshikatsu Yoshii and his colleagues proposed a slightly different model in which the thickness of the lithosphere changes with age. They also assumed that the bottom of the lithosphere is defined by the

Box 2-1. Cooling of the Oceanic Lithosphere

The structure of the oceanic lithosphere changes systematically with age (the time since its creation at mid-ocean ridges), t. The bathymetry (the depth of the ocean floor), d, changes with age as

$$d = d_o + a\sqrt{t},$$ (B2-1-1)

where d_o is the depth at the mid-ocean ridge, and the heat flow, q, changes with age (roughly) as

$$q = q_o + b/\sqrt{t},$$ (B2-1-2)

where a and b are constants. Such relationships can be explained by a simple model of the cooling of the oceanic lithosphere, assuming a concept of convection-induced plate formation at mid-ocean ridges. According to plate tectonics, hot materials rise at mid-ocean ridges and are cooled, and this cold and strong portion (i.e., the oceanic lithosphere) moves laterally. Therefore, the thickness of the oceanic lithosphere increases as they move and get old. The rate at which the lithosphere thickens is controlled by thermal conduction and the distribution of the heat source. If we assume a constant heat flux from the deep mantle and no radiogenic heat generation in the shallow mantle (and crust), then the evolution of the lithosphere is controlled simply by heat conduction. Since the only physical parameter that controls the rate of cooling (of a half-infinite space) is thermal diffusivity, κ, whose dimension is $(m^2 s^{-1})$, the thickness of a cold layer (the lithosphere), h, changes with age as

$$h \propto \sqrt{\kappa t}.$$ (B2-1-3)

As a cool and dense layer develops, it should sink because of the principle of *isostasy*. The principle of isostasy dictates that the pressure at a certain depth must be the same at all points. The physical reason for this principle is that no substantial pressure difference can be supported by materials in the deep portions of Earth because of the relatively low viscosity. For example, for a typical viscosity of $\sim 10^{21}$ Pa·s and a strain-rate of $10^{-15}\ s^{-1}$, the differential stress is ~ 1 MPa. Therefore the pressure difference will not exceed ~ 1 MPa

(continued)

at the depth where viscosity is $\sim 10^{21}$ Pa·s. This stress (~ 1 MPa) is much smaller than typical values of pressure (~ 10 GPa at ~ 300 km depth), so the principle of isostasy is a good approximation. In the context of the evolution of the lithosphere, the depth at which the pressure must be equal is at the base of the lithosphere (i.e., the asthenosphere). Thus, by equating the pressure at this depth at the mid-ocean ridge and at a certain place away from the ridge, one gets

$$\rho_w d_o + \rho_a (d - d_o + h) = \rho_w d + \rho_l h, \tag{B2-1-4}$$

where $\rho_{w,a,l}$ is the density of water, the asthenosphere, and the lithosphere, respectively, hence,

$$d = d_o + \frac{\rho_l - \rho_a}{\rho_a - \rho_w} h, \tag{B2-1-5}$$

leading to equation (B2-1-1).

The heat flow changes with age. The heat flow is controlled by the temperature gradient, $\Delta T / h$ (ΔT is the temperature difference between the asthenosphere and the ocean) and the thermal conductivity, k, by Fourier's law, namely,

$$q = q_o + k \frac{\Delta T}{h}, \tag{B2-1-6}$$

from which equation (B2-1-2) follows.

melting temperature of materials (their model differs slightly from that of McKenzie because Yoshii, Kono, and Ito (1976) incorporated more realistic pressure-dependent melting temperature).

So far we have discussed two different models of evolution of the oceanic lithosphere. One shown in box 2-1 considers cooling of a half-infinite space from above (namely, from the ocean). In this model, we do not care what is going on in the deep portions. Second are the models by McKenzie (1967) or by Yoshii, Kono, and Ito (1976) who considered that the bottom of the plate is where partial melting starts. In fact, a number of textbooks tell us that the lithosphere-asthenosphere structure can be explained by the solidification from a partially molten state; a partially molten region is soft and called the asthenosphere, and its solidification produces a strong lithosphere. Nonetheless, the observational facts presented above can be explained without considering partial melting. The only process needed to explain the above observations is the cooling of materials from above, mainly because the above observations are related to the properties of the lithosphere and not much to those of the asthenosphere. The properties of the asthenosphere are incorporated as a boundary condition at the bottom of the lithosphere, but the thermal structure of the lithosphere depends only weakly on the boundary conditions. Besides, the thermal boundary condition assumed is not consistent with the current knowledge of petrology. The model by Yoshii and others assumes that the asthenosphere is partially molten everywhere and that its solidification over the entire oceanic region corresponds to the evolution of the oceanic plate. Recent petrological studies, however, show that extensive partial melting and solidification take place only near mid-ocean ridges, and that the degree of partial melting for most of the asthenosphere is very small (< 0.1%) if any.

In any case, these models, focused on the properties of the lithosphere alone, cannot place useful constraints on the state of partial melting. To infer the state of partial melting, we need to use observations that are sensitive to the property of the asthenosphere. One example is a recent observation, which I briefly mentioned before, that the boundary between the lithosphere and the asthenosphere is sharp (enough to create seismic reflections) and does not strongly depend on age. In the next section I will explain this observation and, more generally, the relation between partial melting and the asthenosphere.

2-2-2. Partial Melting and the Asthenosphere: The Failure of Classical Models

To understand how partial melting is related to the formation of the lithospheric plate, we have to focus on observations related to the asthenosphere. Whereas the lithosphere can be explained as a cold and strong layer at the surface, anomalous properties of the asthenosphere (low seismic velocity, high attenuation, high electric conductivity, and low viscosity) cannot be explained simply by high temperature. To understand this problem, we first need to consider how the physical properties of rocks, particularly their elastic properties, can be affected by various factors. I will discuss this issue following the historical development.

The most direct way to investigate the influence of various factors on seismic wave velocities would be to measure the elastic wave velocities of minerals and rocks under various conditions in the laboratory, the approach initiated by Francis Birch. The elastic wave velocities of a number of minerals and rocks have been measured as a function of temperature and pressure. Because the sample size in these experiments is small, the frequencies of elastic waves used in the laboratory studies have to be high in order to make wavelengths smaller than the sample size. At these high frequencies (typically 100–1,000 MHz), the effect of *anelasticity*, which I will explain shortly, is very small, and the effect of temperature on elastic wave velocities is mainly due to the reduction in strength of chemical bonding caused by thermal expansion (i.e., the *anharmonic effect* of lattice vibration; see box 1-4). At this frequency range, seismic wave velocities decrease as temperature increases and increase as pressure increases. In Earth's interior, both temperature and pressure increase with depth, but their rates of increase are different (fig. 1-3). While pressure increases almost monotonically with depth, temperature increases rapidly in the shallow regions while increasing very little in the deeper portions. At shallow depths, therefore, the temperature effect dominates, and seismic wave velocities decrease with depth. After some depth, the pressure effect becomes more important, and velocities increase with depth. Seismic wave velocities have a minimum at some depth (≈100 km). As shown in figure 2-3, this model predicts a gradual variation of velocity with depth, so although it can explain the presence of a low-velocity zone around 100 km, we cannot expect seismic reflection from the transition between the high-velocity region (lithosphere) and the low-velocity region (asthenosphere), which is too smooth in this model. The depth variation of velocity must be more drastic to explain seismic observations. What kind of mechanism can cause such a drastic velocity variation?

One possibility is the effect of partial melting expected around this depth. Data used in figure 2-3 are based on measurements at temperatures lower than the melting temperature of mantle minerals, and partial melting could cause an abrupt change in elastic properties. In the paper by Gutenberg, who discovered the low-velocity zone, and in most textbooks, the low-velocity zone (asthenosphere) is explained as a partially molten zone. A common belief still holds: partial melting makes a soft layer, which is the asthenosphere, and the lithosphere (plate) moves above it without much resistance. In fact, explaining low velocity by partial melting is still popular among seismologists.

Hitoshi Mizutani and Hiroo Kanamori, at the University of Tokyo, were the first to test this idea by an experimental approach (1964). A few years later, Hartmut Spetzler and Don Anderson (1968) conducted a similar experiment. These two teams did not use actual mantle rocks but con-

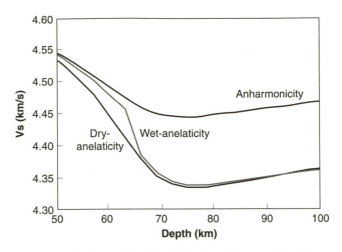

Fig. 2-3. Shear wave velocity as a function of depth for three models. When only anharmonicity is considered, velocity changes with depth rather weakly. Including anelastic effect under "dry" conditions (dry-anelasticity) results in larger depth variation, but still depth variation is smooth. The incorporation of the depth variation in water content (wet-anelasticity) results in a sharp variation in shear wave velocity at ~ 65 km (Karato and Jung 1998).

ducted experiments on analog materials—namely, wood alloy and the ice-salt water system respectively, both of which melt at substantially lower temperatures than do real mantle materials. Their analog experiments demonstrated that a very small amount of partial melt could indeed cause an abrupt reduction in elastic wave velocities, and the hypothesis of a partially molten low-velocity zone seemed to have been supported.

Rick Stocker and Bob Gordon at Yale University, however, cast doubt about this idea (1975). They noted that the physical properties of partially molten materials strongly depend on the melt geometry, and therefore conclusions based on experiments on analog materials might be misleading. They showed that the effects of partial melting are large only when the melt completely wets the grain boundaries. The wetting behavior in turn is controlled by the relative surface energies between the melt and the solid which control the contact angle (dihedral angle) between melt and solid (fig. 2-4). Complete wetting would occur if the dihedral angle is 0°, whereas other geometry would be assumed for higher values of dihedral angles. In fact in many systems, including wood alloy and salt-ice systems, the dihedral angle is 0°. However, it was not known whether the dihedral angle is 0° for the upper-mantle rocks. If the dihedral angle is not 0° then

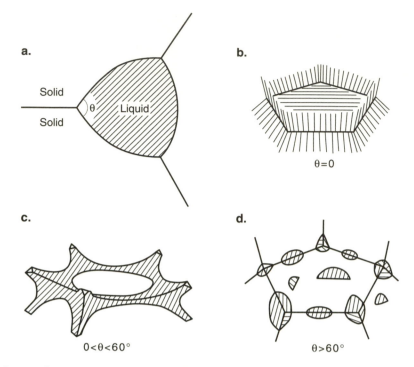

a.

Solid

θ Liquid

Solid

b.

$\theta = 0$

c.

$0 < \theta < 60°$

d.

$\theta > 60°$

Fig. 2-4. The structure of a partially molten medium (a solid-melt mixture). The hatched regions are melt. The shape of the melt strongly depends on the wetting angle between melt and solid. (a) θ is a contact angle between melt and solid and is called the dihedral angle. (b) For the dihedral angle of 0°, melt completely wets the grain boundaries. (c) For $0 < \theta < 60°$, melt assumes continuous tubules along the edges of three grains. (d) For most of mantle materials, $0° < \theta < 60°$, whereas for ice-NaCl or wood alloy, $\theta = 0°$.

the effects of partial melting will be far less than those which have been seen in Wood alloy and salt-ice systems. Following this seminal paper, a number of groups began extensive studies of the morphology of melts in partially molten systems. Harve Waff, at the University of Oregon; Bruce Watson, at Rensselaer Polytechnic Institute; Dave Kohlstedt, at Cornell University, now at the University of Minnesota; and Naoyuki Fujii and Atsushi Toramaru, then at Kobe University, are among those who have participated in this research. These studies have shown that the dihedral angle of basaltic melt with respect to mantle minerals (e.g., olivine) is ~ 30°–50°. Accordingly, the basaltic melt does not completely wet the grain boundaries, but they assume connected tubules along three-grain edged (fig. 2-4). The effect of partial melting is therefore much smaller for upper-mantle rocks, and a drastic reduction in seismic velocity (and other physical properties) is not expected. Subsequent studies by Hiroki Sato and others, then at the Carnegie Institution of Washington, D.C., showed that the effect of partial melting for actual upper-mantle rocks is indeed much smaller than for the wood alloy and salt-ice water systems (Sato et al. 1989).

A similar effect was also found for viscosity (plastic flow). Contrary to the expectation by most people, a systematic experimental study by Dave Kohlstedt's group (1992) showed that the viscosity of olivine-rich aggregates varies only slightly (a factor of 2 to 3) by a small degree (1–3%) of partial melting. Similar to the elastic properties, this result can be explained by a large dihedral angle between melt and solid: because melt does not wet the grain-boundaries in the upper mantle, it can enhance ductile deformation only to a small degree. These experimental findings are not consistent with the classical idea that partial melting significantly reduces seismic wave velocities, causing the low-velocity zone, and reducing the viscosity and facilitating plate motion. One of the paradigms in the theory of plate tectonics is therefore not supported by the experimental studies.

2-2-3. Anelasticity: A Link between Seismology and Rheology

If partial melting does not cause a drastic change in elastic and plastic properties, what is the cause of the asthenosphere? How could we explain a distinct low-velocity and high-attenuation zone in the upper mantle? In this section, I will show that water (or hydrogen) has a key to answer these questions.

Let us first consider the origin of the seismic low-velocity (and high-attenuation) zone—namely, the asthenosphere. An important point to consider when approaching these problems is the effects of *anelasticity*. One might think that a given material has constant elastic wave velocities at a given temperature and pressure irrespective of frequencies of waves, but by conducting accurate measurements, we can find that elastic velocities can vary with frequencies. For lower frequencies, velocities are lower. This is due to anelasticity, the effect of the viscous response of a material on elastic wave propagation (see box 1-1). The effect of viscous deformation is greater for slower deformation, so a medium becomes softer and seismic velocities become lower. Because viscosity is highly sensitive to temperature and other variables, the effect of anelasticity may be able to explain the abrupt variation of seismic velocity. This type of model was proposed by Don Anderson and his colleagues in the early 1980s. To quantitatively test this model, we need to experimentally measure the anelasticity of rocks, which became possible only in the 1990s. The group of Ian Jackson and Mervyn Paterson at Australian National University spent almost ten years developing an apparatus to measure seismic wave attenuation at the frequency range of seismic waves and at high temperatures and high pressures (high pressure is essential for this type of measurements on polycrystalline specimens to suppress the possible effects of cracking at grain-boundaries). They reported the first results of their study on upper mantle rocks in 1992 (Jackson, Paterson, and Fitz Gerald 1992). This truly remarkable achievement in experimental mineral and rock physics provided the first experimental data for understanding the origin of a seismic low-velocity and high-attenuation zone (the asthenosphere).

In order to measure the effect of anelasticity within the range of seismic frequencies, techniques based on wave propagation cannot be used because seismic wavelengths exceed by far the sample size. Instead we have to use quasi-static methods. In this technique, one twists a sample very slowly and observes how a sample deforms. If the anelastic effect is present, there is a slight time lag between applied force and sample deformation, and the amplitude of deformation is smaller than that expected for perfect elasticity. In this kind of experiment, the applied force must be very small. Otherwise sample deformation exhibits nonlinear effects and becomes irrelevant to the interpretation of seismic observations. We therefore need to be able to measure extremely small strains. Ian Jackson's group uses a capacitance transducer for strain measurements, and Hartmut Spetzler's group at the University of Colorado uses a Michelson-Morley-type interferometer. They measure displacement with a resolution of 0.1–1 nm (close to the size of one atom), so these are highly delicate experiments. I once observed this experiment in Hartmut Spetzler's lab. Their laboratory is set up in the basement, and the floor is isolated from the floors of other portions of the building to minimize the vibration. When I started talking, I was very surprised to see that my voice appeared as a big noise on their equipment.

The relation between seismic wave attenuation and seismic wave velocities was formulated by Bernard Minster and Don Anderson at Caltech (1981). When the attenuation of seismic waves occurs following the relation in equation (1-5), then the seismic wave velocities depend on frequency as

$$V(\omega, T) = \overline{V}_\infty(T)[1 - AQ^{-1}(\omega, T)], \tag{2-2}$$

where $V(\omega, T)$ is the seismic velocity as a function of frequency and temperature, A is a constant of order unity related to the frequency dependence of attenuation, and $V_\infty(T)$ is the seismic velocity without attenuation (the anharmonic effect is included here in the $\overline{V}_\infty[T]$ term). This equation shows that seismic velocity decreases by about one to a few percent for a Q of $50 \sim 500$. Equation (2-2) can be also expressed as

$$\frac{\Delta V(\omega, T)}{V_\infty(T)} \equiv \frac{V(\omega, T) - V_\infty(T)}{V_\infty(T)} = AQ^{-1} \propto \eta^{-\beta} \tag{2-3}$$

where the relation in equation (box B1-7-4) is used. Here $\Delta V(\omega, T)$ is the velocity variation due to anelasticity. This relation shows that the anomalies in seismic velocity (and seismic wave attenuation) and viscosity (i.e., rheology) are closely related through the effect of anelasticity.

Using this relation, the depth variation of seismic velocity can be calculated from the temperature dependency of Q, which was determined in the laboratory (equation 1-5). For simplicity, I assume that only temperature (and pressure) can affect anelasticity, and thus seismic velocity, and the result is shown in figure 2-3. The variation of velocity with depth is more marked than the case of where there is no effect of anelasticity, but it is not sharp enough to create seismic reflection. The effect of temperature-dependent anelasticity alone cannot explain the seismic observation.

2-2-4. The Role of Water

Is there anything (other than pressure and temperature) that can abruptly change with depth and have great effects on attenuation and seismic wave velocities? There is one, and it is the concentration of *water* (or hydrogen) contained in mantle minerals. As I will show in this section, even a very small amount of water can considerably affect the viscosity of minerals and their seismic wave attenuation, and the concentration of water in minerals can change abruptly, due to partial melting, with depth. Recent studies have established the effect of water on plastic deformation quantitatively, and it has become possible to discuss the distribution of water in the dynamically evolving Earth's interior. It has been long known that water can be dissolved extensively in the melt but very little in the solid. By examining these results and following a logical argument, one is led to a surprising conclusion: a small degree of partial melting *increases* viscosity and seismic velocity, and the cause of the asthenosphere is the *absence* of (a large degree of) partial melting and the resultant high water content. I will explain next that this seemingly ridiculous conclusion is a consequence of the well-established properties of Earth materials and could provide a natural explanation for a number of observations.

The plastic deformation of rocks is realized by the slow motion of *lattice defects* (such as "vacancies" and "dislocations") in minerals (box 2-2). With some probability, the vibrational energy of atoms is concentrated in these lattice defects, facilitating their movements. As temperature increases, therefore, the viscous flow of rocks becomes more active, and its rate increases exponentially (equation [box B1-7-2]). The motion of lattice defects can also vary greatly with the concentration of impurities. Water (hydrogen) is one of the important impurities, and it has been found that the addition of water increases the concentration of lattice defects and significantly enhances plastic deformation.

David Griggs was a student of Percy Bridgman's at Harvard. After serving in World War II, he joined the faculty at the University of California at Los Angeles (UCLA) in 1948, where he started the experimental study of rock deformation. He developed a new type of high-pressure temperature deformation apparatus and trained a number of students, who are now leading scientists throughout the country. He introduced the modern techniques and concepts of the deformation of solids and also made important contributions to the wide areas of geology and geophysics, including the mechanisms of mountain building and the origin of deep earthquakes. Griggs and his student Jim Blacic were the first to discover the weakening effect of water on the ductile deformation of quartz (Griggs and Blacic 1965) and of olivine (Blacic 1972). However, the study of the origin of the weakening effects of water has a complicated history. Understanding the causes for the water-weakening effects of quartz turned out to be a highly challenging subject; more than thirty years after its discovery, the details of water-weakening effects in quartz remain unclear. The most important reason for this is the fact that the rate of diffusion of water (hydrogen) in quartz is so slow that much of the experimental work has been made under poorly defined conditions with respect to water content (Paterson 1989).

Thanks to the faster diffusion of water (hydrogen), the situation for olivine is clearer. Nevertheless, the history of the study of the water-weakening effect in olivine has interesting twists. Although Blacic (1972) reported a weakening effect of water in olivine (single crystals), his results were not widely accepted until the mid-1980s. Careful studies in several laboratories demonstrated that the apparatus used by Blacic (the Griggs apparatus) produced large errors in stress measurements, which were

Box 2-2. Lattice Defects and Plastic Deformation

The plastic deformation of solids occurs through the motion of *lattice defects*. This concept can be deduced from the calculation of the strength of a perfect material—that is, one that shows a ridiculously high strength. Three physicists, Taylor, Polanyi and Orowan, independently suggested in 1934 that crystal *dislocations* may be responsible for plastic flow in a solid. A key notion here is that deformation occurs much more easily if it occurs by the successive propagation of a "hinge" in a lattice plane rather than by a homogenous motion of two lattice planes. This hinge is called a dislocation. Likewise, Nabarro suggested in 1948 that the diffusion of atoms due to the motion of point defects such as *vacancies* may be responsible for plastic flow at low stresses and high temperatures (diffusion creep). Interestingly, Nabarro pointed out the importance of diffusion creep in Earth's interior in his first paper on diffusion creep. Dislocations and vacancies are collectively referred to as lattice defects, because the perfect periodic array of atoms is disturbed at these "defects." Dislocations are line defects, whereas vacancies are point defects. Both theory and experiments show that the rate of the plastic deformation of solids is proportional to the density and velocity of the motion of these lattice defects. Therefore in general one can write,

$$\dot{\varepsilon} \propto [density\ of\ defects] \times [velocity\ of\ defect\ motion]. \quad (B2\text{-}2\text{-}1)$$

The details of how each term depends on the physical conditions depend on specific mechanisms. The density of dislocations increases with stress. The velocity of dislocations increases with stress and temperature. Therefore the rate of deformation ($\dot{\varepsilon}$) by dislocation motion increases nonlinearly with stress (σ),

$$\dot{\varepsilon} = A\sigma^n \quad (B2\text{-}2\text{-}2)$$

with $n = 3-5$.

In contrast, when point defects are responsible for plastic flow, the rate of deformation is linearly proportional to stress and inversely proportional to grain size (d):

(continued)

$$\dot{\varepsilon} = B \frac{\sigma}{d^m} \qquad\qquad\qquad \text{(B2-2-3)}$$

with $m = 2-3$.

This is due to the fact that the density of point defects does not depend on stress, whereas the velocity of the defect motion (rate of diffusion) is linearly proportional to stress. The driving force for diffusion is the gradient in the concentration of defects at different sides of grain boundaries; hence, the rate of diffusion is inversely proportional to the grain size.

Consequently, dislocation creep (plastic deformation caused by dislocation motion) dominates over diffusion creep at relatively high stresses and for coarse grain sizes. At high temperatures, the motion of dislocation is also closely related to atomic diffusion. Because the rate of atomic diffusion is proportional to the density of point defects, and the density of point defects can be modified dramatically by the addition of impurities, the rate of plastic deformation in solids is often highly sensitive to a small amount of impurities, such as hydrogen.

caused by a high degree of friction. Many scientists became skeptical about Blacic's results. In addition, quartz and olivine have different microscopic deformation mechanisms, so it was not evident why deformation of olivine could be enhanced by water. In quartz, deformation requires cutting the strong Si-O bonds, but in olivine it is not necessary to cut these bonds, a situation that cast further doubt on Blacic's work. In fact, material scientists at the University of Pennsylvania published a paper in 1982, in which they proposed that it was not the effect of water but a change in oxygen fugacity caused by the addition of water that led to the weakening of olivine (Justice et al. 1982).

To settle this issue, Mervyn Paterson's group at Australian National University started a full-scale investigation (Chopra and Paterson 1984; Mackwell, Kohlstedt, and Paterson 1985; Karato, Paterson, and FitzGerald 1986). Paterson, who received his doctorate in physics at Cambridge in 1949 under the supervision of Egon Orowan, who had proposed the concept of crystal dislocation, established a world-class laboratory of rock deformation studies at ANU in the 1950s. Among other contributors, Paterson's high-resolution apparatus for deformation studies has made a fundamental contribution to the rock deformation studies throughout the world. In addition, the extremely careful performance of experimental measurements and of the analysis of the data are characteristic of the research conducted in Paterson's group. I was fortunate to have an opportunity to participate in his group from 1981 to 1984, during which time we were able to finally demonstrate that water indeed enhances the ductile deformation of olivine. When I began this study in 1981, there was confusion about the cause of water weakening in olivine. In addition to Justice and his colleagues (1982), who suggested that this is an apparent effect due to the change in oxygen fugacity by the presence of water, Paterson was arguing strongly for the grain-boundary effects. To deform a polycrystalline aggregate, one needs to deform each grain, but grain-boundary processes must also operate to maintain the continuity of displacement and stress at the grain boundaries. Paterson considered that the main effect of water in enhancing the deformation of olivine aggregates is the enhancement of processes at grain boundaries rather than the enhancement of processes in the olivine grains themselves; hence, water weakening is an indirect effect. This notion was, to me, inconsistent with their own finding that the strength of olivine aggregates under water-rich conditions is much less than that of the strength of single crystals containing no water. Based on this observation, I concluded that it is the weakening of olivine grains themselves that leads to the weakening of olivine aggregates. This point was soon confirmed by the work by Steve Mackwell in the same group (Mackwell, Kohlstedt, and Paterson 1985).

In addition to this intragranular mechanism of water weakening, water weakening may also occur through inter-granular deformation processes. A well-known deformation mechanism in this class is *diffusion creep* (box 2-2), in which plastic deformation occurs through diffusional mass transport between adjacent grain boundaries. Although the importance of this mechanism was suggested by previous studies, reliable quantitative data were not available in the early 1980s. Therefore, I decided to investigate the role of water and grain-size on the plastic deformation of olivine aggregates. The aims of that study were to identify the role of water in intra- and intergranular processes of deformation in olivine aggregates.

A high-precision apparatus developed by Paterson undoubtedly played the key role in this research, but equally important were the careful synthesis of experimental samples and a comprehensive sample analysis (infrared absorption, electron microscope observation, oxygen fugacity measurements, and so on). Grain size must be controlled, oxygen fugacity must be similar between "wet" (water-rich) samples and "dry" (water-poor) samples, and so forth, and the effects of both grain size and water must be characterized separately. Therefore, it took three years to com-

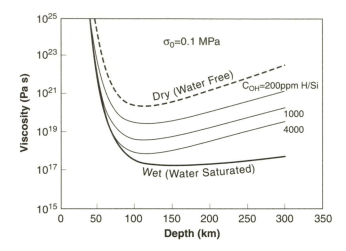

Fig. 2-5. The effect of water on the viscosity of Earth's upper mantle calculated for a typical geotherm (after Karato and Jung 2002).

plete one paper. We conducted deformation experiments of olivine aggregates under almost identical conditions, except for the concentration of water, and showed that creep strength (viscosity) varied dramatically between dry (water-free) and wet (water-saturated) conditions. In addition, we also showed that it was hydrogen dissolved in olivine crystal that enhances deformation: there is a good correlation between the amount of water (hydrogen) dissolved in minerals and the degree of weakening. After this work, this problem has continued to be studied by Dave Kohlstedt's group and my group at the University of Minnesota, and the quantitative relationship between water content and strain rates has been established (fig. 2-5; Mei and Kohlstedt 2000; Karato and Jung 2002). Briefly, the effects of water are large (the strain rate is increased by a factor of ~ 10–100 under typical experimental conditions) and increase as the amount of water dissolved in olivine increases.

Let us consider how water can be dissolved into minerals and change their properties. When we say water in minerals, many people tend to think only of hydrous minerals such as serpentine ($Mg_3Si_2O_5[OH]_4$). A series of studies started by Griggs, however, has shown that a significant amount of water can be dissolved into quartz (SiO_2) and olivine ($[Mg,Fe]_2SiO_4$), whose chemical formulas do not contain OH (these minerals are often referred to as *nominally anhydrous minerals*), and that the dissolved water modifies various properties of minerals. For example, the

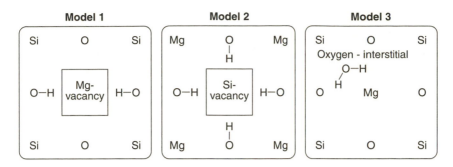

Fig. 2-6. Dissolution mechanisms of water (hydrogen) in olivine. Hydrogen decreases the electric charge of the atomic void, thereby facilitating void generation. The presence of water thus increases the concentration of the atomic void. Atoms move around more easily, and this enhances deformation.

high-pressure experiments performed at Bayreuth Geoinstitut by Dave Kohlstedt and his colleagues showed that the maximum amount of water that can be dissolved in nominally anhydrous minerals in Earth's mantle far exceeds the amount of sea water (Kohlstedt, Keppler, and Rubie 1996). In these nominally anhydrous minerals, water (or hydrogen) is dissolved into their crystal structure as impurity atoms similar to those dissolved in Si (silicon) or Ge (germanium) to control their electrical properties. In olivine $((Mg,Fe)_2SiO_4)$, water may be dissolved through the following mechanisms (fig. 2-6):

1. Oxygen goes into the normal sites for oxygen in olivine. Hydrogen goes into the sites for Mg or Fe.
2. Oxygen goes into the normal sites for oxygen in olivine. Hydrogen goes into the sites for Si.
3. Oxygen goes into interstitial sites. Hydrogen comes next to that oxygen.

For each situation, we can calculate the amount of dissolved water as a function of oxygen fugacity and other factors, and by comparing this figure with experimental results, we can estimate which model is most plausible (Bai and Kohlstedt 1993). These studies show that the dominant mode of dissolution of water (hydrogen) in olivine is the mechanism 1 above (Bai and Kohlstedt 1993; Mei and Kohlstedt 2000; Karato and Jung 2002).

Water (hydrogen) dissolved into minerals as "point defects" affects various properties of minerals. At mantle temperatures and pressures, the concentration of point defects in truly anhydrous olivine is about 1–10 ppm, but the hydrogen solubility reaches 100–10,000 ppm. Hydrogen dissolved as a proton can easily move through the crystal lattice (Mackwell and Kohlstedt 1990), leading to an increase in electric conductivity (Karato 1990). In addition, hydrogen dissolution causes an increase in the concentration of other types of point defects, thus indirectly affecting other physical properties. The analysis of the experimental data on deformation shows that the incorporation of hydrogen increases the concentration of point defects at the Si site in olivine that controls the rate of deformation (Karato and Jung 2002). As a result of enhanced plastic deformation, seismic velocity decreases through the anelastic effect (Karato 1995, 2002).

2-2-5. A New Lithosphere-Asthenosphere Model

Even a small amount of water can significantly enhance the ductile deformation of minerals. On the other hand, a small degree of partial melting only slightly enhances the plastic deformation. Having these two contrasting results at hand I wondered, What would happen if partial melting actually occurs in Earth? How does the effect of water interact with the effect of partial melting? The answer was easy to get, but a surprising one: a small amount of partial melting in a realistic Earth would lead to *hardening* due to the removal of water from olivine. After I returned to Japan, I wrote up this idea and sent it to *Nature* (Karato 1986). The idea of "hardening caused by partial melting" was later elaborated by Greg Hirth, then at the University of Minnesota, who analyzed the details of melting processes and their consequence for rheological stratification (Hirth and Kohlstedt 1996). Karato (1995) and Karato and Jung (1998) examined the consequence of partial melting and the resultant water redistribution for seismic wave velocities and the origin of the seismic low-velocity and high-attenuation zone (asthenosphere).

To clarify the concept behind this idea of hardening caused by partial melting, I need to briefly explain how terrestrial magmatism works. Let us consider mid-ocean ridge magmatism, which is the largest magmatic activity on Earth, as an example. At mid-ocean ridges, mantle materials rise from the deep interior of Earth because of mantle convection. Upwelling leads to a decrease in pressure. The melting temperature of a rock increases with pressure, so as a result of upwelling, the melting point of mantle material decreases. On the other hand, the convective upwelling

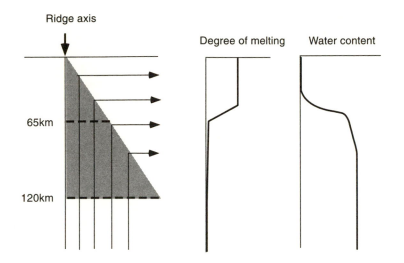

Fig. 2-7. Melting at mid-ocean ridges and the redistribution of water. Mantle upwelling beneath a mid-ocean ridge initially has a small amount of water. Water is contained in olivine and other minerals, and it enhances solid deformation. As mantle upwells, partial melting takes place, and the amount of water contained in the minerals decreases because water is more compatible with melt than with minerals (Karato and Jung 1998). As a result, the viscosity of the mantle increases, as well as its seismic velocity.

rate is much faster than thermal conduction, so the temperature of the upwelling material does not decrease much. Therefore, at some depth, the temperature of the upwelling material exceeds its melting temperature. This is a mechanism for *melting due to adiabatic upwelling* (or pressure-release melting) (fig. 2-7).

To be more precise, melting starts at the depths of about 120–150 km, and at the beginning, only a very small amount of melt ($\approx 0.1\%$) is generated, the amount of which is determined by the original water content. The amount of melt gradually increases with upwelling, and it rapidly increases after the temperature exceeds the starting point for the melting (solidus) of anhydrous mantle material at the depth of about 65 km. Eventually, melt is segregated to form a big stream and erupts on the surface to make a volcano.

This process of igneous rock formation is inferred on the basis of the petrological and geochemical studies of igneous rocks and the seismological study of oceanic mantle. According to this model, the degree of partial melting in the largest portion of the oceanic asthenosphere is small, about 0.1% or less, although this estimate depends on the water content in the mantle. The amount of water contained in Earth's mantle can be estimated from various petrological and geochemical observations and is smaller than the maximum water content that the mantle minerals can dissolve. In other words, most of Earth's mantle is undersaturated with water. It is also important to recall that water is much more soluble in melt than in solid minerals.

By putting together these well-established concepts, we can make the following model. When mantle rocks are at the great depths, a certain amount of water (hydrogen) is dissolved into olivine and other minerals. Because of this, mantle rocks are soft (low viscosity) and their seismic velocity is low. As mantle materials rise, the degree of partial melting increases, and a large fraction of water (hydrogen) is removed from minerals into melt. Consequently, the solid component becomes stronger as partial melting occurs. At this stage, because the amount of melting is small, the bulk property of a rock is determined mostly by the properties of solid minerals. Therefore, the viscosity and seismic wave velocity of mantle materials increase due to the partial melting at this stage. This transition from soft mantle (asthenosphere) to hard mantle (lithosphere) takes place abruptly at around 65 km (for a typical temperature profile under mid-oceanic ridges), where a larger degree of melting begins. After that, when the degree of partial melting exceeds 10% or so, the amount of melt can be large and the effect of partial melting can be significant enough to soften mantle. This softening is, however, limited in the proximity of mid-ocean ridges. Figure 2-8 shows mantle viscosity as a function of depth predicted by this model. At the depth of about 65 km, viscosity suddenly drops, and this viscosity change defines the asthenosphere. At greater depths, the pressure effect on viscosity becomes more dominant, so viscosity increases with depth. A prominent low-viscosity layer exists only when the effect of water is notable. Therefore, we do not expect to have a distinguished asthenosphere in Venus, which is believed to have almost no water. This might be a cause of absence of plate tectonics on Venus, which is believed to have much less water than Earth but to be otherwise very similar.

In this model, abrupt changes in seismic wave velocity and viscosity take place at around 65 km near the mid-ocean ridge, and materials with such a layered structure will passively migrate from the ridge, keeping the layered structure of water content unchanged (the diffusion rate of hydrogen is faster than that of other atomic species, but the diffusion distance is less than a few km for the 100 million years). Therefore the depth

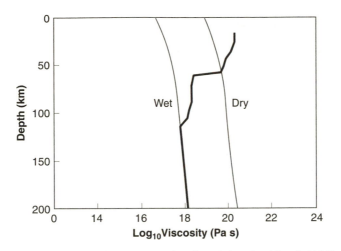

Fig. 2-8. The viscosity profile of oceanic mantle (after Hirth and Kohlstedt 1996).

of the lithosphere-asthenosphere boundary, as defined by an abrupt change in seismic velocity, should not depend on the age of the sea floor in this model. On the other hand, the absolute value of elastic wave velocities, which can be observed by surface waves, should increase with the age of the sea floor because of the decrease in temperature. Therefore, studies of surface waves, which reveal the overall integrated velocities of the upper mantle, should show an increase in the thickness of the lithosphere with the age of the sea floor, whereas body waves of seismic wave reflections, which are sensitive to subtle but sharp changes in structure, show that the depth where an abrupt change in velocity occurs does not change.

In this model, various geophysical anomalies associated with the asthenosphere (low seismic velocity, high electric conductivity, and low viscosity) are caused by a large amount of water (hydrogen) dissolved in mantle minerals in this depth region. A large amount of water (hydrogen) is present in minerals in the asthenosphere *because the amount of partial melting is low*. Note that this is completely opposite to the conventional idea presented in many textbooks, which equates the asthenosphere with a layer of significant partial melting. The lithosphere is a strong layer of the mantle, in this new model, because of the depletion of water due to partial melting at mid-ocean ridges as well as to low temperatures.

2-2-6. The Continental Lithosphere: The Tectosphere

So far I have discussed the lithosphere-asthenosphere structure beneath oceanic regions. The oceanic regions cover approximately 70% of Earth's surface, and the evolution of the oceanic lithosphere is relatively well understood. In contrast, the structure of the continental lithosphere is more complex, and its formation mechanisms are not well known. Seismological and petrological observations provide a starting point for understanding these problems of the continental lithosphere. First of all, seismology tells us that the continental crust is much thicker than the oceanic crust, and that the thickness of the continental crust varies significantly from place to place. From this fact, we can infer that although the formation process of the continental crust is not uniform over different regions, it generally involves a higher degree of partial melting compared to the process of the formation of the oceanic crust. In addition, the upper part of the continental crust is made of granite, so that its genesis should be different from the genesis of the basaltic oceanic crust. For the generation of the continental crust, it is necessary to have the partial melting of mantle with abundant water. On the other hand, the presence of water is not required for the formation of oceanic crust.

From seismological observations, the high-seismic-velocity continental lithosphere is about 200–400 km thick. A truly thick lithosphere (e.g., thicker than 300 km) is, however, limited to old continents (older than ~ 2.7 billions of years, the period called the Archean age). A relatively young continent (formed during the Proterozoic age) such as Eurasia, for example, is at most 200 km thick, but it is still twice as thick as the oceanic lithosphere. What kind of structure does this thick continental lithosphere have? How did it form and how has it survived through billions of years?

Although the idea of a thick continental lithosphere has long been known, it was Tom Jordan, then at the Scripps Oceanographic Institution, who first addressed the geophysical constraints and proposed a petrological model to explain the origin of a thick continental lithosphere. In his first paper on the continental lithosphere (1975), Jordan noted that a very thick continental lithosphere has a dynamical problem. If the high seismic velocity of lithosphere is due to low temperatures, there must be a high-density layer with a 400-km thickness beneath old continents. Then it becomes very difficult to explain why that dense layer can be stable over 3 billion years. Furthermore, if there is a thick, dense layer, it should be detected as a gravity anomaly, which is not observed. Therefore, we must conclude that the continental lithosphere has high seismic velocity but does not have high density (on the other hand, the oceanic lithosphere has both high velocity and high density, so it gradually subsides with age). To explain this, Jordan hypothesized that because the continental crust is formed by a higher degree of partial melting than the oceanic crust, the residual continental lithosphere has a chemical composition slightly different from that of the oceanic lithosphere. Elements such as Al, Ca, and Fe tend to concentrate in the melt (and therefore crust), so these elements have been removed from the continental lithosphere. Consequently, the amount of dense mineral, garnet, which is made of these elements (particularly Al and Ca), is likely small in the continental lithosphere. In addition, a higher degree of partial melting results in the depletion of Fe from the continental lithosphere. Together, the difference in chemical composition between the continental lithosphere and the oceanic upper mantle makes the continental lithosphere less dense than the oceanic upper mantle at the same temperature. To emphasize this compositional difference, Jordan called this continental lithosphere the *tectosphere*.

A problem with this tectosphere hypothesis is why such a thick layer can be stable without notable deformation over 2–3 billion years. By several different methods, we can show that, at the depth of 300–400 km, mantle temperature beneath the continent is about the same as mantle temperature beneath the old ocean floor (Sclater, Parsons, and Jaupart 1981). All other factors being equal, then, mantle viscosity should be similar beneath these two different regions, so it is strange that a thick continental lithosphere (tectosphere) has survived without much deformation. The continental lithosphere, with the age of 3 billion years, must have moved around Earth's surface over its geological history. If the lithosphere is soft and deformable, it would be impossible to keep the thickness of 300–400 km to the present time. The continental lithosphere must be somehow much stronger than the surrounding mantle to maintain its thickness (Doin, Fleitout, and Christensen 1997; de Smet, van den Berg, and Vlaar 1998). Why, then, is it stronger?

In the continental environment, subduction of the oceanic lithosphere often occurs at its edges, which carries water into the continental lithosphere. In fact, the formation of the continental crust requires a high degree of partial melting involving water. Having this notion in mind, several scientists proposed that the rheological structure of the continental upper mantle is determined by that of "wet" (water-rich) olivine (e.g. Kohlstedt, Evans, and Mackwell 1995). However, such a model leads to a weak continental lithosphere, which would not survive for billions of years.

Henry Pollack, at the University of Michigan, proposed that the continental lithosphere has maintained its integrity throughout geological history due to the increase in viscosity caused by the removal of water by partial melting (Pollack 1986). In other words, Pollack assumed that the continental lithosphere is "dry" (water-poor). The concept of a "dry" continental lithosphere is somewhat paradoxical. The formation of the continental crust requires partial melting in the presence of water, whereas the formation of the oceanic crust does not require water. Therefore, some regions of the continental upper mantle must have a higher water content than does the oceanic upper mantle. So why does a higher water content in the continental upper mantle lead to a higher degree of depletion of water from the continental lithosphere? The answer can be found in the effects of water on the melting relationship of mantle materials. When a large amount of water is present, the melting starts at a greater depth and the degree of partial melting will be larger. The partial melting leads to the depletion of water from minerals. Therefore, in regions where the original water content is high, water depletion occurs to a greater depth through partial melting, resulting in a thicker lithosphere. This idea is consistent with the results of numerical modeling showing that water content (as well as temperature) in Earth's mantle may have decreased with time. McGovern and Schubert (1989) showed that the water content in Earth's mantle has been declining with time due to continuous degassing (removal of a volatile component by volcanism). Therefore the water content in early Earth is expected to have been higher than it is today. Consequently, the partial melting in early Earth would have occurred at greater depths and resulted in a thicker continental lithosphere. This notion is quite consistent with the seismological observations indicating that the continental lithosphere in the Archean age (older than ~ 2.7 billion years) is considerably thicker than the lithosphere of Proterozoic continents.

How can we test this hypothesis? One way is to measure the electric conductivity of the continental lithosphere. Electric conductivity depends not only on temperature but also on the amount of water (hydrogen) dissolved in minerals (Karato 1990). Therefore, if the temperature of the continental lithosphere can be reasonably well estimated, we can estimate the amount of water using measured electric conductivity. Hirth, Evans, and Chave (2000) suggested that the low electrical conductivity in the old (Archean) continental lithosphere is consistent with a low water content.

In short, the rheological structure of the continental upper mantle is strongly affected by the distribution of water and is likely to be highly heterogeneous. The continental upper mantle near subduction zones likely contains a large amount of water and is weak, whereas much of the continental lithosphere is depleted with water and is strong.

Finally, a few remarks on the continental crust may be made. The thickness of the continental crust varies a great deal from one place to another (Mooney et al. 1998). In the western United States (the Basin and Range region, which includes New Mexico and Nevada), the crust is less than ~ 20 km thick, whereas in areas in the collision zone, such as Tibet, it is ~ 50–70 km thick. The causes for this regional variation in crustal thickness include the difference in volcanic activity and in the degree of tectonic thickening and thinning: the collision of two continents tends to thicken the crust, whereas the crust tends to be thinned in regions of extension. The crust is often divided into two sublayers: upper and lower crust. The continental upper crust is made of silica-rich rocks such as granite, whereas the lower crust is made of more Mg, and Fe-rich rocks, such as basalt or some metamorphic rocks formed from basalt or andesite (Rudnick and Fountain 1995). Rocks with basaltic chemical composition change to a dense rock called eclogite at the depth of ~ 40–50 km. The density of eclogite is larger than that of upper mantle peridotites. Consequently, crustal materials transformed to eclogite are gravitationally unstable and tend to be removed from the shallow regions. Therefore, if the continental crust (made mainly of materials with basaltic composition) reaches a certain thickness, its deeper portion may be removed and sink to even deeper portions. These dense materials (eclogite), however, would not go to the bottom of the mantle. Estimates of the densities of various materials based on high-pressure experiments show that eclogite (or more garnet-rich materials) is denser than other ambient materials in the upper mantle and the transition zone, whereas it is less dense than the ambient materials in the shallow lower mantle (see chap. 4). Consequently, if this removal (delamination) of the lower continental crust occurs, these materials should be accumulated in the transition zone, leading to the enrichment of relatively silica-rich (and less depleted) components in the transition zone. There is some evidence showing the absence of the lower crust in some regions, suggesting that this crustal delamination occurs in some regions.

THREE • SEISMIC TOMOGRAPHY AND MANTLE CONVECTION

3-1. SEISMIC TOMOGRAPHY

Seismic tomography revolutionized the way in which we study the dynamics of Earth's interior. Before the advent of seismic tomography, studies on the dynamics of Earth were based on observations of near-surface regions, such as the dynamics of plate motions; and the dynamics of the deep interior were treated by mere speculation, without observational constraints. With seismic tomography, we have obtained, for the first time, strong observational constraints on which we can develop models of the dynamics of Earth's deep interior.

Tomography is an "inverse" method by which one infers the invisible three-dimensional structure of some object (in our case Earth) from the data obtained outside of it. This technique was first developed in medical science to infer fine structural variations in the human body by the use of X-ray absorption.

In seismic tomography, one investigates three-dimensional small deviations of real Earth structure from a standard Earth model. In standard Earth models such as PREM (the Preliminary Reference Earth Model; see chap. 1), the physical properties of Earth's interior are considered to vary only as a function of depth (radially symmetric structure); seismic velocity does not depend on azimuth and polarization (i.e., the structure is isotropic), except in the uppermost mantle, where a simplified anisotropic structure is assumed. This is of course an idealized Earth model, and the real Earth deviates slightly from this standard model. By estimating these deviations, we can construct three-dimensional models of the structure of Earth. Although the deviations are small (a few percent in most cases), the model contains valuable information regarding the dynamic Earth and its evolution. Thermal and chemical heterogeneities can be expected for the dynamically deforming and evolving interior, and seismic anisotropy may be generated by structural anisotropy that results from the deformation of rocks by convection.

It is not straightforward, however, to interpret the results of seismic tomography and other high-resolution seismic studies in terms of dynamics. Without sufficient understanding of mineral physics, not only is this valuable seismological information useless; it can also be misused. In this chapter, I summarize the recent results of seismic tomography as well as those of high-resolution seismology. I then present the essence of mineral and rock physics, at the atomic level, which is required for interpreting the results of these seismological studies and for understanding their implications for dynamics of Earth's interior.

3-2. HETEROGENEITY

3-2-1. Seismological Observations

Most of seismic tomographic studies determine the fine deviations of the structure of real Earth from a standard Earth model, assuming that seismic wave velocities are isotropic. Whereas velocities in a standard model depend only on depth, actual velocities can vary from place to place, even at the same depth. This is called heterogeneity. We first note that the assumption of isotropic velocity is not always valid. Because, in some cases, the degree of anisotropy can be as strong as the degree of heterogeneity derived by isotropic inversion, heterogeneity could be (incorrectly) inferred from anisotropy. To avoid this, seismic velocities must be measured in various directions, and their azimuthal average must be taken.

There are various methods to measure heterogeneities in seismic velocities. The most direct one is to use a number of travel times of seismic waves passing through the same region to determine the seismic velocity structure of the region. This approach was first employed independently by Keiiti Aki (then at Massachusetts Institute of Technology) and Adam Dziewonski (at Harvard) in the late 1970s (Aki, Christoffesson, and Husebye 1977; Dziewonski, Hager, and O'Connell 1977). Applications to surface waves and free oscillations were developed by the Caltech group (Toshiro Tanimoto, Ichiro Nakanishi, and Don Anderson), the University of Paris group (Jean-Paul Montagner), and John Woodhouse (then at Harvard University and now at Oxford University). Groups at the University of California at Berkeley (Barbara Romanowicz), in San Diego (Guy Masters), in Austin, Texas (Steve Grand), and at Tohoku University, Japan (Akira Hasegawa and Dapeng Zhao), have also made major contributions.

In these studies, in addition to the study of travel time inversions, a more elaborate method using seismic waveforms is often employed (box 3-1). Guust Nolet, then at University of Utrecht, now at Princeton University, and his colleagues in the Netherlands as well as John Woodhouse have made fundamental contributions in this area. In the waveform inversion, whole seismic records are used in determining the best-fit model. In contrast, in the travel-time method, only the (first) arrival times are used in the modeling. In inferring large-scale heterogeneities, the fine spectral structure of the free oscillation of the whole Earth is also used, because the existence of heterogeneity can slightly shift the periods of free oscillation and split peaks in the spectrum (this is analogous to the effects of fine scratches of a flute that modify the tone slightly). By studying free oscillations, one can detect not only heterogeneity in seismic velocity, but also heterogeneity in density, because some modes of free oscillation involve a radial motion of materials in the gravitational field. In fact, the first recognition of global-scale heterogeneities came from the analysis of free oscillation data (Masters, Jordan, Silver, and Gilbert 1982). In 1984, global-scale upper- and lower-mantle heterogeneities were mapped separately, using surface waves and body waves, respectively (Woodhouse and Dziewonski 1984; Dziewonski 1984). These studies were the first attempts at whole mantle seismic tomography and had a great impact on the study of mantle dynamics. Brad Hager and his colleagues (Hager et al. 1985) immediately incorporated the results of these seismic tomographic images into convection models and discussed possible mantle flow patterns and corresponding variations in the gravity field. Assuming that velocity heterogeneity resolved by seismic tomography was caused by thermal heterogeneity, they converted velocity anomalies to density anomalies, from which they calculated mantle flow. We will later discuss the validity of this important assumption in more details.

Since then, the resolution of seismic tomography has rapidly improved, owing to progress in theory, increasing computational capability, and the installment of digital seismometer networks. Until late 1980s, the resolution of tomography was so limited that the scale of detectable heterogeneity was as large as a few thousand kilometers. Even with this low resolution, however, both high- and low-velocity anomalies were detected in the lowermost mantle (D″ layer), and these anomalies were inferred to correspond to subducted materials and the source regions of hotspots, respectively.

In 1990s, scientists around the world conducted high-resolution tomographic studies. In the early 1990s, Toshiro Tanimoto, then at Caltech,

Box 3-1. Principles of Seismic Tomography

In a standard Earth model, the structure of Earth changes only with depth. However, in real Earth, lateral variations in seismic wave velocities (and densities) may occur. When these variations occur, their signatures can be found in the seismographs. Consider a case shown below (fig. B3-1-1), where a region of anomalous velocity occurs in the central portion (shown by thick hatching). Seismic waves X1-C and X2-B pass through this region and show some signatures of this anomaly, whereas other seismic waves such as X1-A, X2-A, and X3-C do not show the effects of this anomaly.

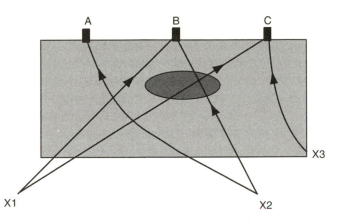

Fig. B3-1-1. A schematic diagram showing the principle of seismic tomography. A shaded region has "anomalous" seismic velocities. X1, X2, and X3 are the sources for seismic waves (earthquakes), and A, B, and C are seismic stations. The rays X1-B, X2-A, and X3-C do not pass through anomalous regions, so the seismic records show normal, expected results. In contrast, rays such as X1-C or X2-B pass through anomalous regions and show anomalies in seismic records, including travel times and waveforms.

In seismic tomography, one seeks evidence of these anomalous regions in various aspects of seismic records. A typical seismic record (called a seismogram) is shown in fig. B3-2. The simplest approach to detect these anomalies is to look at the travel times of some representative waves (such as the P-wave or S-wave). If this anomalous region contains materials with velocities higher than background materials, then travel times of seismic waves in the seismograms for X1-C and X2-B will be faster than expected from a standard model. However,

(continued)

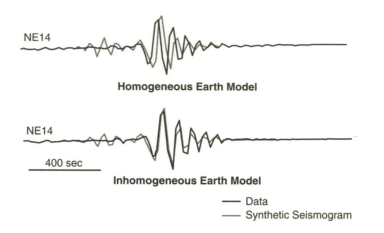

Fig. B3-1-2. An example of seismic waveform inversion (Nolet 1987). The top Fig. compares the observed waveform with the synthetic waveform calculated from a homogeneous model. The bottom figure shows the results for an inhomogeneous model.

travel times are only a small fraction of all seismic records that one has. Much richer information is contained in the whole seismogram. The seismic tomographic method, in which whole record (the shape of the seismogram) is used, is referred to as *waveform inversion* and provides a high-resolution picture of velocity anomalies (fig. B3-1-2).

and his colleagues published the results of global-scale upper-mantle tomography using surface waves (Zhang and Tanimoto 1991, 1993; Anderson, Tanimoto, and Zhang 1992). Their study revealed, for the first time, the difference between mid-ocean ridges and hotspot volcanoes as well as the deep structure of continents on a global scale. About at the same time, Yoshio Fukao, then at Nagoya University in Japan, and Rob van der Hilst, then at University of Utrecht in the Netherlands, investigated the fate of subducted slabs in the deep mantle using body-wave tomography, and they exposed the complex interaction between subducted slabs and ambient mantle in the transition zone (van der Hilst et al. 1991, 1997; Fukao et al. 1992). Steve Grand (1994) and Rob van der Hilst and his colleagues (van der Hilst, Widiyantoro, and Engdahl 1997) showed evidence of the penetration of slabs to the lower mantle.

It is more difficult to detect hot upwelling flow, compared to cold subducted slabs, because body waves tend to avoid low-velocity regions corresponding to hot upwellings. Although surface waves do not suffer from this biased sampling, it is still difficult to detect a narrow low-velocity region, such as the root of a plume, because surface waves have a limited spatial resolution. Nevertheless, mantle directly beneath Hawaii and Iceland has been investigated intensively by a sophisticated waveform inversion method, and the source regions of these "hotspot" volcanoes are beginning to be revealed (Wolfe et al. 1997; Bijwaard et al. 1998; Foulger et al. 2001; Zhao 2001).

Recently, Miaki Ishii and Jeroen Tromp, at Harvard (1999), derived a model for whole mantle density heterogeneity from free oscillation data. Briefly, they used the distortion of peaks of several modes of free oscillation of Earth that are sensitive to density distribution. Some modes of free oscillation involve the radial motion of materials against gravity. Consequently, the frequencies (and the form of peaks) of these modes are sensitive to the density. However, inferring density anomalies from seismic records is challenging work because the signature of density anomalies is subtle. Because of these technical difficulties, the validity of the results of Ishii and Tromp is still debated. Nonetheless, the information on the density anomalies is highly critical to the understanding of the origin of heterogeneities, and the gross feature of density anomalies inferred by Ishii and Tromp is in harmony with other independent studies of joint velocity inversion, such as one by Masters and his colleagues (2000).

Observational constraints on heterogeneity can be summarized as follows (fig. 3-1).

1. There is a large degree of heterogeneity around 100–200 km depth, which has a strong correlation with surface tectonics. Seismic velocities are slow beneath mid-ocean ridges and fast beneath old continents. Low-velocity zones beneath volcanoes near mid-ocean ridges, however, do not continue straight down to greater depths and often show a large bend. These low-velocity zones are also limited to shallow depths (< 150 km). Based on this observation, Toshiro Tanimoto and his colleagues argued that magmatism at mid-ocean ridges is "passive." That is, mid-ocean ridge magmatism is not driven by hot upwelling from deep mantle, but the relative motion of plates creates a void at the surface, which then causes the upwelling of mantle materials and resultant melting. These results have been confirmed by a later high-resolution tomographic study by Jean-Paul Montagner and Jeroen Ritsema (2001) (see fig. 3-2).

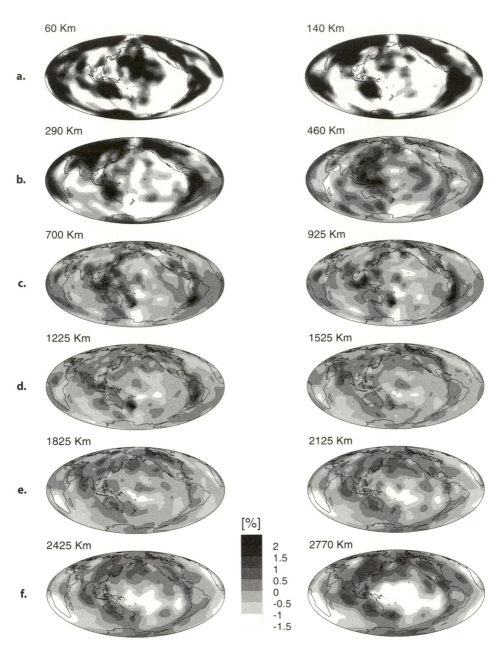

Fig. 3-1.

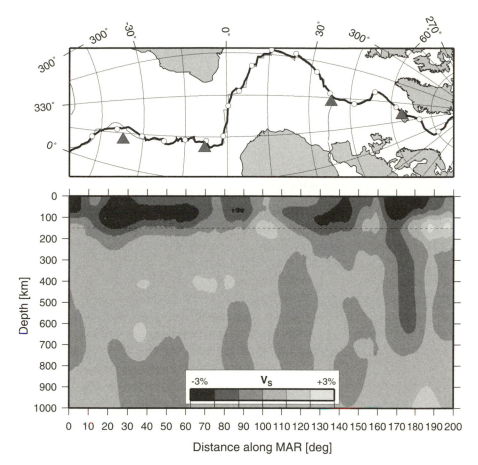

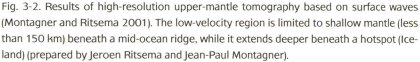

Fig. 3-2. Results of high-resolution upper-mantle tomography based on surface waves (Montagner and Ritsema 2001). The low-velocity region is limited to shallow mantle (less than 150 km) beneath a mid-ocean ridge, while it extends deeper beneath a hotspot (Iceland) (prepared by Jeroen Ritsema and Jean-Paul Montagner).

Fig. 3-1. An example of whole-mantle seismic tomography (after Masters et al., 2000). The deviation of shear wave velocity from a reference model is shown at different depths. (a) The amplitude of velocity anomaly generally decreases with depth, but it increases in the lowermost mantle. (b) In the lowermost mantle, there are extensive low-velocity anomalies beneath Africa and the south Pacific. (c) In the transition zone (460 km), there is a wide region of high-velocity anomaly beneath the western Pacific. (d) Beneath the circum-Pacific regions, there is a high-velocity anomaly in the lowermost mantle. (e) A velocity anomaly at a very shallow level (60 km) is closely related to surface tectonics. It is slow beneath mid-ocean ridges and fast beneath continents. (f) High-velocity anomalies beneath continents extend to the depth of 200 – 300 km while low-velocity anomalies beneath oceans become blurred deeper than 200 km (prepared by Gabi Laske and Guy Masters).

2. The correlation between velocity anomalies and surface tectonics decreases with depth. The amplitude of heterogeneity also decreases with depth.

3. At the lowermost mantle (2,300–2,900 km depth), however, the amplitude of heterogeneity becomes larger. There is some degree of correlation, though rather crude, between the pattern of heterogeneity in this region and surface tectonics. For example, there is a fast velocity anomaly beneath the circum-Pacific region. On the other hand, large, low S-wave velocity anomalies can be seen beneath Africa and the south Pacific. There are ultra-low-velocity anomalies (more than 20–30%), mostly within these low-velocity regions (Garnero et al. 1998). Partial melting and chemical heterogeneity (e.g., Fe rich) are among possible causes. The ultra-low-velocity regions may be the source region for hotspots located above them, but they may also be caused by the sedimentation of light materials from the core below (see chap. 6; Buffett et al. 2000). The analysis of free oscillations by Ishii and Tromp (1999) shows that the lowermost mantle beneath large, low S-wave velocity anomalies such as Hawaii are denser than average. Similarly, Masters and his colleagues (2000) showed that the bulk sound wave velocity (V_ϕ; $V_\phi^2 \equiv V_p^2 - \frac{4}{3}V_s^2$) has positive anomalies in these regions where S-wave velocity has negative anomalies.

4. The amplitude of compressional wave (P-wave) velocity anomalies is about half that of shear wave (S-wave) velocity anomalies and generally increases with depth (Masters et al. 2000) (fig. 3-3).

5. The nature of heterogeneity at mid-mantle depths (500–1,000 km) is rather controversial, though its amplitude seems to be larger than that at adjacent depths. Guy Masters and his colleagues at Scripps Oceanographic Institution (Masters, Jordan, Silver, and Gilbert 1982), who were the first to study global-scale seismic velocity anomalies, concluded that there were long-wavelength anomalies in the mantle transition zone. Studies by other researchers, including Yoshio Fukao (Fukao et al. 1992) and Rob van der Hilst (van der Hilst et al. 1991) suggest that mid-mantle velocity anomalies are highly spatially variable. Beneath the western Pacific, high-velocity anomalies of a large horizontal extent are observed at many places above (or below for a few locations) the 660-km discontinuity. For these regions, subducting plates do not seem to penetrate easily into the lower mantle, and it appears that they bend and stagnate within the transition zone, at least temporally. At most of the subduction zones in the eastern Pacific (particularly around Central America), on the other hand, this type of a velocity anomaly is not observed, suggesting that plates are

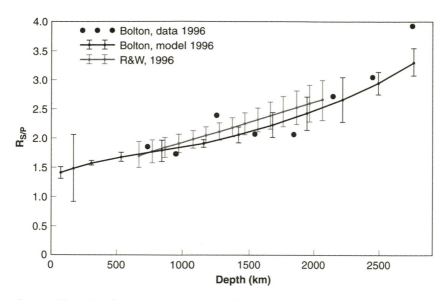

Fig. 3-3. The ratio of shear-wave heterogeneity to compressional-wave heterogeneity, $R_{s/p}$ (after Masters et al. 2000; Bolton 1996; Robertson and Woodhouse 1996).

subducting into the lower mantle without much difficulty. The origin of the difference in the styles of subduction is an important problem, which I will examine in detail in chapter 4.

3-2-2. The Significance of Lateral Heterogeneity in Geodynamics: Fundamentals of Mineral Physics

The next step after mapping velocity anomalies is to examine their implications for geodynamics. Though slow- and fast-velocity anomalies are usually considered to represent high- and low-temperature anomalies, respectively, it is actually not so obvious whether this interpretation is valid, even in a qualitative sense. In this section, I will discuss the origins of seismic velocity anomalies from the microscopic viewpoint of mineral physics to better illustrate the geodynamic significance of lateral heterogeneity in seismic wave velocities and densities.

The lateral variation in seismic wave velocities (and density) should be due to the lateral variation of the physical environment (temperature, pressure etc.) and/or chemical composition. Among other physical environments, heterogeneity of pressure cannot be large because of the low strength of Earth materials in the deep portions (remember the principle of isostasy). Therefore, the possible causes for lateral variation are variations in temperature and/or chemical composition. In a dynamically evolving planet like Earth, one expects that both temperature and chemical composition should change laterally; therefore, if one can map temperature and/or chemical heterogeneity, one would have important constraints within which to understand the dynamics and evolution of Earth. If velocity anomalies are mainly caused by temperature variations, a fast-velocity anomaly means lower temperatures, where cold and dense materials may be sinking. If velocity anomalies are caused by variations in the amount of iron, however, a fast-velocity anomaly indicates lower iron concentration, which corresponds to a lower-density anomaly. These regions may be rising. Velocity anomalies, therefore, have completely different dynamical implications depending on whether their origin is thermal or chemical.

A. The Effects of Temperature

Let us first consider how temperature anomalies affect velocity anomalies. There are a number of possibilities for chemical heterogeneity, so it is difficult to derive a unique conclusion if we try to interpret velocity anomalies as chemical signatures. Therefore, it would be appropriate to first consider thermal effects as the simplest explanation, and then if thermal effects cannot explain the observations, we may consider other possibilities, including chemical effects.

The effects of temperature on elastic properties include the *anharmonic effect* (due to thermal expansion) and the *anelastic effect*, as explained in chapter 2. It is important to distinguish these two effects here. The variation of seismic velocities due to temperature can be measured in the laboratory, and results for representative minerals are shown in table 3-1. These results are based on ultrasonic waves or opto-elastic techniques (techniques using the interaction of photons with phonons), so the velocities are measured at 100–1,000 MHz frequencies; hence, the results correspond to the anharmonic effect by thermal expansion. Some of these results have been known since the mid-1960s and have been used to infer the temperature dependence of seismic wave velocities. However, if we actually use these results, we will face three difficulties.

First of all, very extensive partial melting would be predicted throughout the mantle. About ± 5% of the velocity variation seen in the shallow upper mantle corresponds to a temperature variation of ± 600 K, if the temperature derivative for olivine (the representative mineral in the upper

TABLE 3-1

Temperature Derivatives of Seismic Wave Velocities for Typical Minerals
Determined at High Frequencies and Corresponding Values of $R_{s/p}$

	$-\dfrac{\partial \log V_p}{\partial T}$ $(10^{-5}\ K^{-1})$	$-\dfrac{\partial \log V_s}{\partial T}$ $(10^{-5}\ K^{-1})$	$R_{s/p}$
MgO	7.2	9.4	1.30
CaO	6.8	8.4	1.23
grossular	4.6	6.0	1.30
pyrope	4.6	4.5	0.98
olivine (Fe/(Fe+Mg)=0)	6.2	7.6	1.22
olivine (Fe/(Fe+Mg)=0.1)	7.4	7.7	1.04
Al_2O_3	4.7	6.9	1.47

mantle) shown in table 3-1 is used. As explained in chapter 1, the average temperature at the shallow upper mantle is already close to the melting temperature. Thus, if low-velocity regions are hotter than normal mantle by 600 K, then, an extensive degree of melting is expected there. This contradicts geochemical observations as well as other geophysical observations (e.g., shear wave propagation through low-velocity regions), which indicate that extensive partial melting (exceeding a few percent) is limited to the very vicinity of volcanoes.

The second difficulty can be seen in the ratio of compressional to shear wave velocities. This ratio,

$$R_{s/p} \equiv \frac{\delta \log V_s}{\delta \log V_p}, \tag{3-1}$$

is sensitive to the mechanism of velocity anomalies, so it is one of the important parameters from which one could identify the origin of seismic heterogeneity. If the change in seismic wave velocities is due to temperature, then equation (3-1) can be written as

$$R_{s/p} = \frac{(\frac{\partial \log V_s}{\partial T})}{(\frac{\partial \log V_p}{\partial T})}. \tag{3-2}$$

When temperature derivatives come through the anharmonic effect, the results of high-frequency measurements can be used to evaluate $R_{s/p}$ using the values in table 3-1. The results show that $R_{s/p} \sim 1.0–1.5$, which is significantly smaller than the values obtained from seismic tomography.

The third problem regards the density anomalies estimated from velocity anomalies. Consider the ratio of velocity and density anomalies,

$$R_{\rho/s,p} \equiv \frac{\delta \log \rho}{\delta \log V_{s,p}}. \tag{3-3}$$

Again, assuming thermal origin, this parameter can be calculated from the temperature dependence of density (i.e., thermal expansion) and of the seismic wave velocities listed in table 3-1. The results show that $R_{\rho/s,p} = 0.7–1.0$ for most materials. In contrast, a comparison of seismic tomographic observations and the geodynamic interpretation of gravity shows $R_{\rho/s,p} = 0.2–0.4$ in most regions and sometimes even negative values in the very deep lower mantle (fig. 3-3). Thus, the density anomalies within the Earth are much smaller than those predicted from this simple model.

Among these three problems, the inconsistency with the observed ratio of shear to compressional wave anomalies, $R_{s/p}$, is generally considered to be serious because the observational basis for $R_{s/p}$ is robust. Based on this discrepancy between the observed value and the experimental results, many scientists have proposed that the majority of seismic velocity anomalies indicate chemical heterogeneity (Robertson and Woodhouse 1996; Kennett et al. 1998; Masters et al. 2000). It is too hasty, however, to abandon the notion of a thermal origin on the basis of experiments conducted at high frequencies under low pressures. Seismic waves propagate through high-temperature and high-pressure regions in the Earth's interior, and their frequencies are much lower than those used in laboratory experiments. Experimental results from laboratories obtained at high frequencies are not necessarily applicable to the interpretation of subtle changes in seismic waves. In addition, the temperature dependence of elastic wave velocities in solids may be different under extreme compression under deep Earth conditions.

Before explaining this problem further in detail, let us review some basics of mineral physics. First of all, why does elastic wave velocity decrease when the temperature is raised? Equation (1-1) shows that elastic wave velocity depends on density and elastic constants. An increase in temperature leads to thermal expansion, so density decreases. As a result, the distance between neighboring atoms increases, leading to weaker atomic bonding. Anharmonicity refers to the effect caused by the change in atomic bonding due to thermal expansion. Because the change in atomic bonding usually has a greater effect than the change in density caused by a temperature increase, elastic (seismic) wave velocity decreases with increasing temperature. To see this point in more detail, let us use Birch's law on seismic wave velocities. Based on a large number of measurements, Francis Birch derived an empirical law in elasticity (1961): the variation of elastic wave velocity with temperature and pressure occurs mainly through the variation of density. In other words, even under different temperature and pressure conditions, the elastic wave velocity of a material should stay constant as long as its density does not change. Using Birch's law, one can show that $R_{s/p}$ is nearly unity if velocity variation is due to temperature variation through *anharmonic effects*. The mineral physics considerations summarized in box 3-2 show that this ratio is related to another fundamental material parameter, called the Grüneisen parameter, as

$$R_{s/p} = \frac{\gamma_s - 1/3}{\gamma_p - 1/3} \tag{3-2}$$

where $\gamma_{s,p}$ is the Grüneisen parameter corresponding to the S- and P-wave, respectively. According to Grüneisen, a German physicist, these parameters are nearly constant ($\sim 1.0 - 1.5$) for a wide range of materials under various conditions. Consequently one expects $R_{s/p}$ to be nearly unity (see table 3-1). Therefore, the observed values of $R_{s/p} = 2.0 - 3.5$ in the mantle are hard to reconcile with a thermal effect involving anharmonicity.

We can derive one more important conclusion from mineral physics. Using the definition of the Grüneisen parameter and the definition of thermal expansion, we get

$$\left(\frac{\partial \log V_{s,p}}{\partial T} \right)_{ah} = -\alpha_{th} (\gamma_{s,p} - 1/3). \tag{3-3}$$

Now the coefficient of thermal expansion, α_{th}, decreases significantly with pressure (box 3-3): the value of thermal expansion at the bottom of the mantle is $\sim 1/3$ of the value at the surface. The Grüneisen parameter changes only weakly with depth. Consequently, $\left| \left(\dfrac{\partial \log V_{s,p}}{\partial T} \right)_{ah} \right|$ significantly decreases with depth, and hence the amplitude of velocity anomalies corresponding to the same temperature variation should decrease significantly with depth. Therefore, a significant decrease in the amplitude of velocity anomalies with depth does not imply a decrease in tempera-

Box 3-2. The Thermal Origin of Velocity Heterogeneity Due to Anharmonicity

Lateral variation in temperature is a possible cause of lateral variation in seismic wave velocities (and density). Temperature variation causes variation in seismic wave velocities due to two different atomistic mechanisms. One is the *anharmonicity* of lattice vibration—that is, the change in strength of interatomic potential due to *thermal expansion*. When this is the cause of lateral variation in velocities, we can write

$$\delta \log V_{s,p} = (\frac{\partial \log V_{s,p}}{\partial T})_{ah} \delta T, \tag{B3-2-1}$$

where δT is the the lateral variation in temperature and the suffix *ah* is used to emphasize that the quantity is evaluated for anharmonicity only. Now, according to *Birch's law*, the variation in seismic wave velocity with temperature is through the variation in density. Therefore,

$$(\frac{\partial \log V_{s,p}}{\partial T})_{ah} = \frac{\partial \log V_{s,p}}{\partial \log \rho} \frac{\partial \log \rho}{\partial T} = -\frac{\partial \log V_{s,p}}{\partial \log \rho} \alpha_{th}, \tag{B3-2-2}$$

where α_{th} is the thermal expansion coefficient defined by

$$\alpha_{th} \equiv -\frac{\partial \log \rho}{\partial T}. \tag{B3-2-3}$$

Next, consider the term $(\frac{\partial \log V_{s,p}}{\partial \log \rho})$ from the microscopic viewpoint. The seismic wave velocity is determined by the strength of the interatomic potential as follows. Let us assume a simple periodic potential (see box 1-4),

$$\phi_{s,p} = A_{s,p}[1 - \cos(2\pi \frac{x}{a})], \tag{B3-2-4}$$

where x is the distance in a lattice line and a is the interatomic spacing. This potential causes a force when an atom moves from its stable position. This force is related to the elastic constant relevant to S- and P-waves ($\mu_{s,p}$)—namely, $\mu_{s,p}(x/a) = \phi'_{s,p}/a^2$, where prime (') indicates the derivative with respect to x. Therefore,

$$A_{s,p} = \frac{\mu_{s,p} a^3}{(2\pi)^2}. \tag{B3-2-5}$$

Now, this atomic potential can also causes atomic vibration with a frequency of $\omega_{s,p} = \sqrt{\phi_{s,p}'' / m}$. Thus,

$$\omega_{s,p} = \sqrt{\frac{\mu_{s,p} a}{m}}. \tag{B3-2-6}$$

Now, remembering that seismic wave velocity is given by $V_{s,p} = \sqrt{\frac{\mu_{s,p}}{\rho}} = \sqrt{\frac{\mu_{s,p} \cdot a^3}{m}}$, we have

$$\omega_{s,p} = V_{s,p}/a. \tag{B3-2-7}$$

By differentiating (B3-1-7) with respect to density, we get

$$\frac{\partial log V_{s,p}}{\partial log\rho} = \frac{\partial log\omega_{s,p}}{\partial log\rho} - \frac{1}{3}. \tag{B3-2-8}$$

Now, let us define a Grüneisen parameter:

$$\gamma_{s,p} \equiv \frac{\partial log\omega_{s,p}}{\partial log\rho}. \tag{B3-2-9}$$

Grüneisen showed that this parameter takes nearly constant values, 1.0 ~ 1.5, irrespective of materials. From equations (B3-2-2), (B3-2-8), and (B3-2-9), it can be readily shown that

$$\left(\frac{\partial log V_{s,p}}{\partial T}\right)_{ah} = -(\gamma_{s,p} - \frac{1}{3})\alpha_{th} \tag{B3-2-10}$$

and

$$R_{s/p} = \frac{\gamma_s - 1/3}{\gamma_p - 1/3}. \tag{B3-2-11}$$

ture anomalies with depth. The results can be attributed to a nearly constant amplitude of lateral variation in temperature variation.

Is it possible to explain the observed value of $R_{s/p}$ ($\approx 2-3.5$) by modifying the theory of anharmonicity? Several scientists have tried to explain the results of seismic tomography through pressure effects on the Grüneisen parameter. Although the Grüneisen parameter is nearly constant, it does change slightly with pressure, and using quantum mechanical (first-principles) calculations, one can show that this ratio, $R_{s/p}$, changes from ~ 1.5 at the shallow part of the mantle to ~ 2.0 at the bottom of the mantle (Isaak et al. 1992; Karato and Karki 2001). Therefore,

Box 3-3. The Pressure Dependence of Thermal Expansion

Let us begin with the definition of thermal expansion, $\alpha_{th} \equiv -\dfrac{\partial log\rho}{\partial T}$ (B3-2-3). By taking the derivatives of this equation with respect to pressure, one can show that

$$\frac{\partial log\alpha_{th}}{\partial log\rho} = \frac{1}{\alpha_{th}}\frac{\partial logK}{\partial T} \equiv \delta_T. \qquad \text{(B3-3-1)}$$

Orson Anderson showed that the parameter, δ_T, (the Anderson-Grüneisen parameter), is nearly independent of pressure and temperature and takes values of $\sim 4-6$ for most materials. In this case, equation (B3-3-1) can be integrated to yield

$$\frac{\alpha_{th}(\rho)}{\alpha_{th}(\rho_o)} = \left(\frac{\rho}{\rho_o}\right)^{-\delta_T}. \qquad \text{(B3-3-2)}$$

We can see from equation (B3-3-2) that thermal expansion decreases significantly with pressure (depth).

this partly solves the problem with $R_{s/p}$, but three problems remain: (1) although the pressure effects on anharmonicity increases $R_{s/p}$, the observed values of $R_{s/p}$ are still larger than the prediction based on anharmonicity; (2) the temperature anomalies estimated from velocity anomalies are too large for the shallow mantle; and (3) the density anomalies predicted from velocity anomalies are too large compared to the observations.

Most of these three problems can be solved simultaneously by invoking a significant contribution of *anelasticity* (Karato 1993b; Karato and Karki 2001). The basic idea behind this model is quite simple. The point is that at the frequency range of seismic waves, the anelastic effect caused by viscous flow, in addition to the anharmonic effect, is also important. This increases the temperature sensitivity of seismic wave velocities, so a large variation in seismic wave velocities could be attributed to a relatively small temperature variation. This effect (the effect of anelasticity) is larger for S-waves than for P-waves. Thus, large values of $R_{s/p}$ can be attributed to the role of anelasticity. Finally, changes in seismic wave velocities through this mechanism occur without much change in density (unlike the effects of anharmonicity). Therefore, it leads to a small $R_{\rho/s,p}$. Though the importance of anelasticity has been recognized since the early 1970s, as already mentioned in chapter 2, only relatively recently has anelasticity been found to also be important for the interpretation of seismic tomography.

The starting point is equation (B1-6-1) for the temperature dependence of anelasticity (Q) and (2-2) for the effects of anelasticity on seismic wave velocities. Seismic wave velocity depends on Q, and Q is very sensitive to temperature. Therefore, attenuation (or anelasticity) significantly affects the temperature dependence of seismic wave velocity. The anelastic effect is larger for shear waves than for compressional waves, so the greater degree of shear wave heterogeneity can be explained. Because of the anelastic effect, a small increase in temperature can cause a large drop in velocity, but not in density. This can explain small values of $R_{\rho/s,p}$. Let us examine these results more quantitatively in the following.

By differentiating (2-2) by temperature and using (B1-6-1), we obtain (assuming $Q \gg 1$),

$$\frac{\partial log V_{s,p}(\omega, T)}{\partial T} = (\frac{\partial log V_{s,p}}{\partial T})_{ah} + (\frac{\partial log V_{s,p}}{\partial T})_{an}, \tag{3-4}$$

where the first and second terms of this equation express the variation of seismic wave velocity caused by temperature variation through the anharmonic and anelastic effects, respectively. The second term can be calculated using equation (B1-6-1) and can be calculated from the seismologically determined values of Q and the experimentally determined value of the temperature dependence of Q.

The effect of anelasticity on the parameters $R_{s/p}$ and $R_{\rho/s,p}$ can be calculated from equation (3-4). The results are shown in figure 3-4. Due to the paucity of experimental data on deep mantle materials, this type of discussion has large uncertainties. We can, however, still conclude the following:

1. Anelasticity significantly increases the temperature dependency of seismic wave velocity. This effect is significant and almost doubles the temperature derivative of velocity. Thus, we can explain velocity anomalies in the upper mantle with reasonable temperature anomalies.
2. The value of $R_{s/p}$ increases because of anelasticity. By also considering the pressure dependence of anharmonicity, $R_{\rho/s,p}$ increases from 1.5 to 2.0–2.7.
3. Anelasticity lowers the value of $R_{\rho/s,p}$. This can be calculated based on mineral physics using a seismologically measured Q. $R_{\rho/s,p}$ is 0.2–0.4 with the effect of attenuation considered.

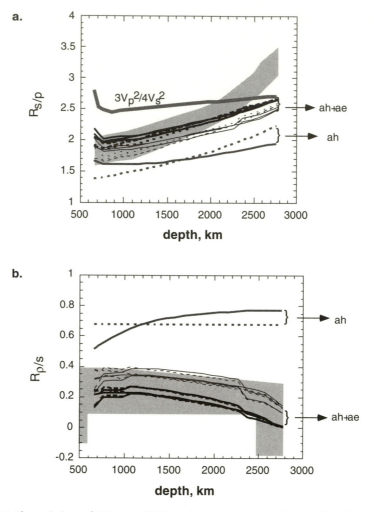

Fig. 3-4. The variations of (a) $R_{s/p}$ and (b) $R_{p/s}$ due to temperature effects only (after Karato and Karki 2001). By considering anelasticity as well as anharmonicity, results consistent with most of the observations can be obtained (Karato 1993b). Lines show the results of mineral physics assuming the thermal origin of velocity and density heterogeneity. *ah* is the anharmonic effect only, and *ah* + *an*, both the anharmonic and anelastic effects. A thick solid line indicates the upper limit of $R_{s/p}$ for thermal effects. Shaded regions show seismological/geodynamical estimations.

Most of the three problems discussed previously, therefore, can be explained, by including the effect of anelasticity, as having a thermal origin. However, seismic observations deeper than ~ 2,300 km are still difficult to explain. The values of $R_{s/p}$ there (2.7–3.5) are too large for a thermal origin. Furthermore, $R_{\rho/s,p}$ becomes negative according to the recent study of Ishii and Tromp (1999). These observations indicate the presence of chemical heterogeneity in this deep region. Similarly, structural differences between continental and oceanic mantle require different chemical compositions (chap. 2). Note that this conclusion is similar to the one by van der Hilst and Karason (1999), who argued for the presence of a thick, chemically distinct layer in the lower mantle (~ 1,600 to 2,900 km depth). However, Karato and Karki (2001) showed that chemical anomalies are required only in the very deep regions (D″ layer) of the lower mantle.

B. The Effects of Heterogeneity in Chemical Composition

B-1. The Effects of Major Element Chemistry

Because there are a number of possible types of compositional anomalies, drawing general conclusions for chemical heterogeneity is not a straightforward process. Our discussion here is limited, therefore, to a few representative examples. As already seen for the case of continental tectosphere, differences in the degree of partial melting are the most obvious cause for chemical heterogeneity. Earth evolves chemically as a result of continuous partial melting. It is quite natural to expect chemical heterogeneity in the evolving Earth. While all concentration ratios change as a result of partial melting, let us focus, for simplicity, on the ratio of iron (Fe) and magnesium (Mg). Because Fe selectively goes to melt, the concentration of iron is lower for regions that have experienced a greater degree of partial melting. This applies not only to low pressures but also to high pressures in the lower mantle.

The variations of density and elastic properties as a function of the iron-magnesium ratio (Fe/[Fe + Mg]) are studied for many minerals, as summarized in table 3-2. When iron content increases, density increases sharply, whereas elastic constants vary little or even decrease. Thus, when the iron-magnesium ratio varies, the ratio of density to velocity variations, $R_{\rho/s,p}$, is negative with a large absolute value. On the other hand, an increase in the iron content reduces seismic wave velocity, with a larger effect on shear wave velocity. As a result, $R_{s/p}$ becomes greater than unity

TABLE 3-2

Effects of Change in Fe/(Fe + Mg) on Velocity and Density Variation Ratios $(R_{s/p}, R_{p/s})$

	$R_{s/p}$	$R_{p/s}$
(Mg,Fe)O	1.27	−1.04
pyrope (Mg,Fe)$_3$Al$_2$Si$_3$O$_{12}$	1.22	−0.71
olivine (Mg,Fe)$_2$SiO$_4$	1.40	−0.96

by variations in the iron content. However, the value of $R_{s/p}$ caused by the variation in the iron-magnesium ratio is still too small compared to the observed values. Similar calculations may be made for the silicon-magnesium ratio, and the results are similar. Therefore, we can conclude that changes in the concentration of these elements alone cannot explain the observed anomalies in the very deep portions of the lower mantle. Karato and Karki (2001) suggested that the change in calcium content might be responsible for the anomalies in the very deep lower mantle.

By combining the effects of temperature and chemistry involving calcium, therefore, we may be able to explain the large $R_{s/p}$ and the very small (negative) $R_{p/s,p}$ observed in the lowermost mantle (the D″ layer). Recently, Ishii and Tromp (1999) estimated density and velocity anomalies in the mantle simultaneously by analyzing free oscillation data, and they found an anticorrelation between density and velocity anomalies below ∼ 2,500 km. This indicates that velocity anomalies at these depths are caused by chemical heterogeneity.

Lateral variation in chemical composition and temperature can cause lateral variation in stable phases. Even at the same depth, the combination of stable phases can be different for different temperatures and chemical compositions, so density and elastic properties can greatly differ. An example is the differences between continental and oceanic mantle, as mentioned above. Continental mantle is depleted in Al and Ca, so it is expected to be less abundant with dense garnet than the oceanic mantle is.

B-2. The Effects of Water (Hydrogen)

Hydrogen is a minor chemical species that deserves special attention. We have already discussed its effects on anelasticity and its resultant influence on seismic wave velocities. Unlike other major elements, (a small amount of) hydrogen significantly affects neither seismic wave velocities nor density through anharmonic effects. However, its effects through anelasticity could be significant, especially in regions where attenuation is rather high. These are regions of hot upwelling (mantle beneath mid-ocean ridges or hotspots) or mantle above the subducting lithosphere ("mantle wedge"). Low-velocity regions are likely to be due to high water content as well as high temperature anomalies. Regions of high water content have been identified in the upper mantle beneath Japan (Karato 2002). As will be discussed later in this chapter, water (hydrogen) also has important effects on seismic anisotropy.

To sum up our discussion so far, we may draw the following conclusions:

1. Most of velocity anomalies in the mantle can be explained as temperature anomalies. The effect of anelasticity can explain, for the most part, the large $R_{s/p}$ and small $R_{\rho/s,p}$, which have previously been considered to indicate chemical heterogeneity.
2. However, very large velocity anomalies observed in the shallow mantle (ocean-continent heterogeneity) and in the lowermost mantle (the D″ layer) are difficult to explain by the effect of temperature only.

If seismic heterogeneity is caused by compositional anomalies or different combinations of stable phases, the interpretation of velocity anomalies in terms of mantle dynamics becomes different from the one assuming a thermal origin. Fast velocity anomalies do not necessarily mean cold and dense regions. They may even correspond to lighter regions. Large velocity anomalies are detected at the uppermost and lowermost parts of the mantle (fig. 3-1), and the chemical origin of velocity anomalies is likely at both places, in addition to the thermal origin. The presence of chemically distinct regions in the top and the bottom regions of the mantle is a natural consequence of the chemical differentiation of Earth. Lighter or heavier than normal materials will rise or sink to these regions and will remain there to cause lateral variation in chemical composition. Similarly, the mid-mantle boundary at ~ 660 km might also trap chemically distinct materials, leading to chemical heterogeneity just above and below this boundary. Detecting these anomalies and identifying their origin are critical to the better understanding of the evolution of this planet.

3-3. ANISOTROPY

Real Earth shows another type of deviation from an idealistic model. In our previous discussions we talked about P-wave or S-wave velocities. Such a discussion implicitly assumes that Earth is isotropic. This is not strictly true in most regions of Earth. Real materials can be *anisotropic*. Imagine a piece of crystal such as quartz and measure its elastic wave velocities. It will turn out that the velocities that you measure depend on the orientation of the wave propagation. Also, even for the same propagation direction, you may find two different S-waves with different velocities (box 3-4). These are both manifestations of elastic anisotropy. Of course real Earth is not made of single crystals, but for several reasons, highly deformed portions of Earth would show anisotropic structures. When initially isotropic rock is deformed to large strains, minerals in that rock tend to align their orientations in certain directions. Alternatively, when a rock is made of two distinct materials, then large deformation often results in a layered structure (fig. 3-5). Rocks that contain these structures show elastic anisotropy. In these cases, the nature of anisotropy is determined by the nature of deformation, particularly the geometry of deformation. Therefore, if one can infer anisotropy of Earth's structure, one can learn a lot about the deformation geometry in Earth's interior. For these reasons, a number of studies have been performed to reveal anisotropic structures in Earth to learn more about the dynamics of Earth's deep interior. In this section, after a brief discussion of basic concepts of seismic anisotropy, I will summarize major results of seismic anisotropy with spe-

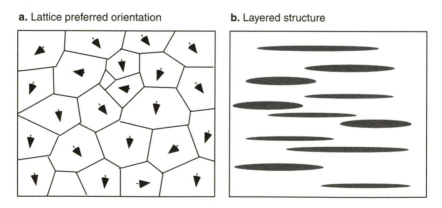

Fig. 3-5. Two types of anisotropic structures: (a) lattice preferred orientation (LPO) and (b) layered structure.

Box 3-4. Types of Anisotropy

There are two types of seismic waves: P-waves (compressional waves) and S-waves (shear waves) (fig. B3-4-1). The particle motion in a P-wave is parallel to the propagation direction, whereas in an S-wave, it is perpendicular to the propagation direction. In this sense, the S-wave is similar to light (an electromagnetic wave). Consequently, the velocity of both P- and S-waves can be dependent on the orientation of wave propagation, which is referred to as *azimuthal anisotropy*. For S-waves, because particle motion in a plane perpendicular to the propagation direction involves a different distortion of materials dependent upon the direction of particle motion, S-wave velocity can depend on the direction of particle motion—that is, polarization. This is referred to as *polarization anisotropy*. When a seismic wave passing through an anisotropy layer

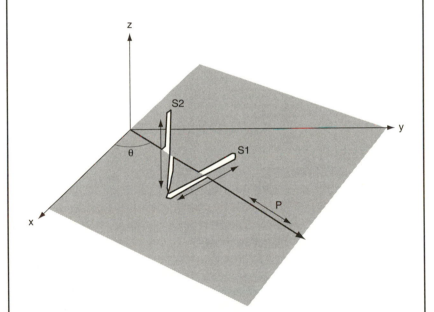

Fig. B3-4-1. A schematic diagram showing the anisotropic propagation and polarization of seismic waves. For a compressional (P-) wave, only azimuthal anisotropy can be observed. For shear waves, both azimuthal and polarization anisotropy can be observed (two shear waves with different polarization, but the same direction of propagation can have different velocities).

(continued)

is recorded at one station, one would find two different S-wave arrivals with different polarizations, rather than one S-wave. This is called *shear wave splitting*. A record of shear wave splitting contains two pieces of information: the travel time difference and the direction of the polarization of the (faster) S-waves. They reflect the strength of anisotropy and the geometry of anisotropic structure, respectively.

cial emphasis on anisotropy in Earth's mantle (anisotropy of the inner core will be discussed in chapter 6).

The elastic properties of isotropic materials are characterized by two elastic constants, and therefore there are only two elastic wave velocities (Vs and $\overline{V}p$). In contrast, in an anisotropic material, a large number of elastic constants are needed to characterize its elastic properties. In an extreme case of the most general anisotropy, we need 21 elastic constants. It is difficult to determine 21 elastic constants for a given region from seismic observations. In many cases, therefore, we assume weak anisotropy to obtain approximate solutions and investigate the characteristics of anisotropy based on them. Also, some simplifying assumptions are often made as to the symmetry of anisotropy. A commonly used model assumes that anisotropy depends only on vertical and horizontal directions. In this model (this type of anisotropy is called *transverse isotropy*), for example, there are two kinds of shear wave velocities, V_{SH} (for waves whose particle motion is in the horizontal plane) and V_{SV} (for waves whose particle motion is in the vertical plane).

3-3-1. Seismological Observations

Anisotropy is a more complex concept compared to heterogeneity. It denotes the directional dependency of velocity. We must first understand two different meanings of "direction." As already explained, seismic waves have two types of waves: (P-) and shear (S-) waves. For compressional waves, the direction of wave propagation is parallel to that of particle displacement (oscillation). For shear waves, these two directions are perpendicular to each other (box 3-4). Thus, similar to light, shear wave velocity can be different for different oscillation directions. Seismic waves, therefore, generally have two types of anisotropy; anisotropy depending on the direction of wave propagation is called *azimuthal anisotropy*, and anisotropy depending on the direction of particle oscillation is called *polarization anisotropy*.

In this chapter we will focus on anisotropy in the mantle (anisotropy in the inner core will be discussed in chapter 6), although the crust also has anisotropy. The seismological study of anisotropy in the mantle was started in the 1960s. Harry Hess, then at Princeton University, who advocated the concept of sea floor spreading, predicted the seismic anisotropy of the oceanic lithosphere (1964). To test this hypothesis, a seismic experiment using explosives was conducted in the eastern Pacific (Raitt et al. 1969). By examining the flow pattern of mantle materials expected from sea floor spreading at mid-ocean ridges, Hess predicted that seismic wave velocity in the mantle should have anisotropy because the crystallographic orientation of a strongly anisotropic mineral, olivine, is likely to be aligned by mantle flow. In this seismic experiment, azimuthal anisotropy was observed for compressional waves propagating through the uppermost mantle. The researchers detonated explosives at the sea surface and measured, using a number of ships, the velocity of compressional waves through the uppermost mantle as a function of the propagation azimuth (fig. 3-6). Although this type of anisotropy is conceptu-

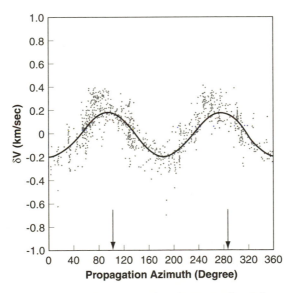

Fig. 3-6. The first observation of upper-mantle anisotropy (after Raitt et al. 1969). The graph shows the compressional wave velocity propagating through the uppermost mantle beneath the eastern Pacific as a function of the propagation azimuth. The fastest azimuths (shown by arrows) are parallel to the direction of plate motion.

ally easy to understand, this type of measurement comes with a number of caveats. Besides practical issues such as costly operations, a serious scientific problem with this method is the difficulty distinguishing anisotropy from heterogeneity. It is not clear that the azimuthal variation of measured seismic velocity is due to anisotropy or heterogeneity, because each velocity measurement depends on a different path of seismic wave propagation. This is serious because the magnitude of anisotropy (less than a few percent) is similar to that of heterogeneity, although large heterogeneity is not expected in many of the oceanic regions.

In this regard, the measurement of polarization anisotropy is generally more reliable. The influence of heterogeneity is almost none because waves with the same propagation path are analyzed. In addition, the measurement can be done with a few seismometers, so it is less expensive. In fact, one of the fast measurements of anisotropy is the polarization anisotropy. Keiiti Aki recognized the systematic difference in velocities of two types of surface waves involving V_{SH} (Love wave) and V_{SV} (Rayleigh wave): in the upper mantle of oceanic region, $V_{SH} > V_{SV}$. Aki proposed a layer of horizontally aligned melt pockets (1968). In this very early stage of the study of anisotropy, important models for anisotropy such as lattice preferred orientation of minerals and a layered structure had already been proposed.

The study of seismic anisotropy was considerably advanced in the 1980s. Masataka Ando, then at Kyoto University, was among the first to develop a powerful method to determine polarization anisotropy using only one seismological station (Ando, Ishikawa, and Wada 1980). Because shear waves with different polarization directions propagate through an anisotropic material with different velocities, these two kinds of shear waves are recorded separately in a seismogram. This is called *shear wave splitting* (box 3-4). Whereas this method provides unequivocal evidence for an anisotropic structure (because the measured difference in velocity is hardly caused by heterogeneity), it is usually difficult to locate the anisotropic structure. For shear wave splitting using the SKS phase (which first propagates through the mantle as a shear wave, then converts to a compressional wave in the core, and then converts to a shear wave again in the mantle (Figure 1-1)); for example, the anisotropic region can be anywhere between the observation point (at the surface) and the core-mantle boundary. Because of its feasibility, however, the study of shear wave splitting has become popular, and polarization anisotropy has been measured at various places in the world (Silver 1996). In certain cases, however, the location of anisotropic structure can be identified through the study of shear wave splitting. For example, when one uses an S-wave that is created by conversion from a P-wave at a discontinuity (such as the 410-km discontinuity), then the polarization anisotropy of the S-wave must come from the region above or below the discontinuity, depending on the direction of wave propagation. Such a technique is useful for locating the source of anisotropy (e.g., Park and Levin 2002).

Despite the uncertainty about the depth of anisotropy, many people believe that an anisotropic region is located in the upper mantle because anisotropy in the deeper regions is rather low as inferred from other data sets. With this assumption, the flow pattern in the upper mantle can be inferred from the measurement of shear wave splitting. The results in major oceanic regions are relatively simple: the direction of polarization of the faster S-wave is nearly parallel to the motion of the lithosphere. The observations in the regions behind subduction zones are more complicated. In many regions, the direction of the fast S-wave polarization is nearly parallel to the trenches, whereas the data far from the trenches show the fast directions at a high angle to the trenches (Buttles and Olson 1998; Fischer et al. 1998). Near hotspots such as Hawaii, the pattern of shear wave splitting follows the flow pattern controlled by the interaction of the plume with the plate-tectonic flow (Walker, Bokelmann, and Klemperer 2001).

In the continental regions, the direction of fast S-wave polarization is usually parallel to the direction of orogenic belts (Silver 1996). This is true not only for recently active regions (e.g., the Himalayas), but also for ancient regions with the last geological activity as long as 2 billion years ago (central North America). By generalizing this result, Silver argues that anisotropy observed in continental regions mostly reflects past geological activities, with little influence from current mantle flow. Some researchers, however, do not agree with this. On the basis of anisotropy measurements at Africa, Lev Vinnik, at the Russian Academy at Moscow, proposed that most of anisotropy is caused by the current mantle flow (Vinnik, Green, and Nicolaysen 1995). By examining global data sets, Atsuki Kubo and Yoshihiro Hiramatsu, in Japan, suggested that anisotropy is formed in the asthenosphere beneath continents (1998). This problem is related to how continents deform on the geological time scale. If there is 300 to 400 km thick, strong tectosphere beneath continents, as Jordan suggested, the anisotropy of continental mantle mostly reflects past geological activities. If deep continental roots are relatively soft and easily deformable, on the other hand, the anisotropy of deep continental mantle may reflect current mantle flow. However, in many regions, more complicated patterns are found and often interpreted to be due to layered anisotropic structure (depth-dependent anisotropy) (Savage and Silver 1993). I will come back to this problem when I discuss the role of water in seismic anisotropy.

From the 1980s, the study of anisotropy using surface waves also made remarkable progress. Toshiro Tanimoto, then at Caltech, played a central role in this contribution. In particular, a paper by Tanimoto and Anderson (1984) with an attractive title, "Mapping Mantle Convection," called many people's attention to the problem of anisotropy. This and subsequent papers by the Caltech group and by Jean-Paul Montagner, at the University of Paris, conducted the global-scale analysis of upper mantle anisotropy (Montagner and Tanimoto, 1990, 1991). These surface wave studies investigated both azimuthal and polarization anisotropy. With surface waves, the depth dependence of anisotropy can be estimated from the wavelength dependence of anisotropy, because a longer-wavelength surface wave is sensitive to deeper structure. As an extension to this kind of effort, Montagner and Kennett (1996) estimated whole-mantle anisotropy from the analysis of free oscillation data (fig. 3-7). Anisotropic structures can be inferred from the free oscillation spectrum through the detailed analysis of the splitting of peaks.

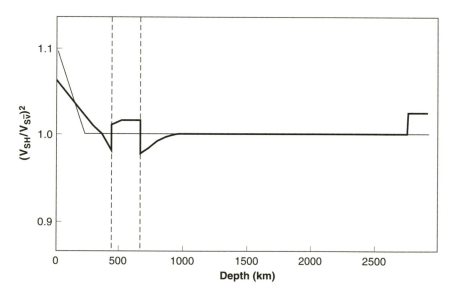

Fig. 3-7. A whole-mantle seismic anisotropy model (after Montagner and Kennett 1996). This result is based on the analysis of free oscillation data with the assumption of anisotropy with the vertical axis of symmetry.

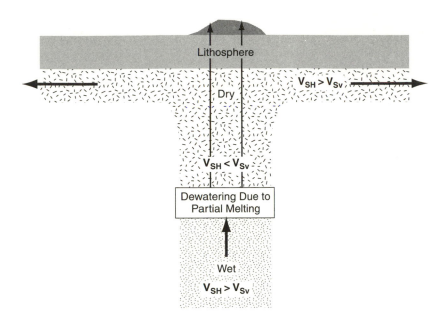

Fig. 3-8. A model for seismic anisotropy beneath hotspots. Plume materials are originally enriched with water and move vertically in the deep portion, causing the $V_{SH} > V_{SV}$ anisotropy. At a certain depth (~ 100 km), significant partial melting occurs, leading to dewatering (removal of water from olivine) (Karato 1986). This changes the anisotropy to $V_{SV} > V_{SH}$. When the plume hits the bottom of the lithosphere, flow direction changes to horizontal. This horizontal flow of "dry" (water-poor) materials causes the $V_{SH} > V_{SV}$ anisotropy.

Another notable result from these surface wave studies is the detection of regional variation in V_{SH}/V_{SV} anisotropy. The Caltech group found that although $V_{SH} > V_{SV}$ in most parts of oceanic upper mantle, the upper mantle beneath mid-ocean ridge shows $V_{SH} < V_{SV}$ (Nataf, Nakanishi, and Anderson 1986). Montagner and Guillot (2000) also found that beneath Hawaii, there is a strong $V_{SH} > V_{SV}$ anomaly. Similarly Gaherty (2001) found that beneath Iceland, anisotropy changes from $V_{SH} > V_{SV}$ in the region deeper than ~ 100 km to $V_{SV} > V_{SH}$ in the shallower region. The global mapping of surface wave anisotropy by Göran Ekström and Adam Dziewonski (1999) also showed strong $V_{SH} > V_{SV}$ anisotropy in the broad region of central Pacific. These observations contain important information on the flow pattern and chemical composition (particularly water) associated with upwelling plumes (fig. 3-8).

Global scale studies show that the anisotropy of deep mantle is much smaller than that of shallow upper mantle (less than 200 km depth). In the lower mantle, almost no anisotropy has been detected except in the lowermost region (the D″ layer). As described later, most mantle minerals have strong anisotropy. Therefore, almost isotropic deep mantle cannot be taken for granted.

This unexpected result needs some explanation in terms of the physics of the deformation of materials and the dynamics of mantle convection. I will come back to this point later in this book. Some authors reported weak anisotropy for the mantle transition zone (Montagner and Kennett 1996), and for the shallow portions of the lower mantle (Wookey, Kendall, and Barruol 2002), but generally anisotropy in this depth range is not well constrained.

The lowermost part of the mantle can be studied in detail by various kinds of waves. Due to large contrasts in elastic properties and densities at this boundary, many waves, including diffracted waves, propagate along this boundary. Therefore, much richer data are available to investigate this dynamically important region (see Gurnis et al. 1998). Thorn Lay and Ed Garnero, at the University of California at Santa Cruz, as well as Mike Kendall and Paul Silver conducted such studies (Kendall and Silver 1996; Lay, Williams, and Garnero, 1998). It is becoming clear that some of the D″ layer has strong anisotropy but other regions do not. Furthermore, the type of anisotropy (i.e., which polarization is faster) also

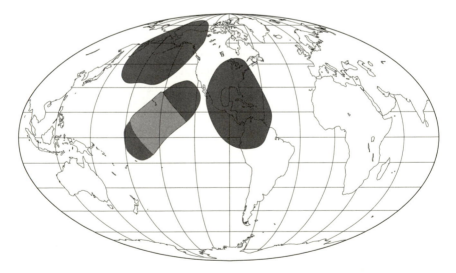

Fig. 3-9. Seismic anisotropy at the bottom of the lower mantle (D″ layer) (after Lay et al. 1998). Regions of strong $V_{SH} > V_{SV}$ are shown by dark hatches. They occur mostly in the circum-Pacific. Anisotropy is highly variable near the center of the Pacific (light gray region).

varies significantly from place to place (fig. 3-9). The anisotropy of the D″ layer may provide important clues for the formation of plumes as well as the interaction between subducted slabs and the core-mantle boundary (Lay, Williams, and Garnero 1998).

The discovery of strong anisotropy in the inner core was totally unexpected. It was first reported in two papers published in 1986 by a group of seismologists at Harvard (Woodhouse, Giardini, and Li 1986; Morelli, Dziewonski, and Woodhouse 1986; see Song 1997 for a review). Geomagnetism is generated in the core, so the study of its fine structure, including anisotropy, has a possible impact on the study of the geodynamo. Inner core anisotropy revealed another surprising finding: the inner core seems to rotate faster than the mantle (Song and Richards 1996; Su, Dziewonski, and Jeanloz 1996). These discoveries in succession have a big impact on the study of core dynamics. Chapter 6 will discuss in detail the anisotropy of the core and its rotation.

3-3-2. The Mineral and Rock Physics of Anisotropy

Once anisotropy is seismologically detected, the next step is to examine its geodynamic significance. For this purpose, we have to understand the mechanisms of the formation of the anisotropic structure. There are two kinds of anisotropic structures (fig. 3-5): the preferred orientation of crystal and the layering of different elastic materials. In the following, we will call them simply *preferred orientation* and *layered structure,* respectively. Preferred orientation refers to the state where the orientations of rock-forming crystals are aligned along particular directions. If these minerals are elastically anisotropic, rocks as a whole also have anisotropy. When a seismic wave whose wavelength is much larger than the thickness of each layer passes through a layered structure, the material would look like a homogeneous but anisotropic body. In this case, the strength of anisotropy depends on differences between the elastic properties of each layer.

Layered structure and preferred orientation have different characteristics in anisotropic behavior. Consider layered structure first. For simplicity, we assume that each layer itself is isotropic. Though only lateral layering is considered below, the following discussion can be also applied to vertical layering. A laterally layered structure has axial symmetry around the vertical axis. The elastic constants of this type of material can be easily calculated (box 3-5), and the main results are summarized as follows:

Box 3-5. The Anisotropy of a Layered Structure

Consider a material with a layered structure, composed of two layers with different elastic moduli, M_1 and M_2, with fraction of thickness d_1 and d_2, respectively ($d_1 + d_2 = 1$) (fig. B3-5-1). Such a material has an anisotropic structure that has a symmetry plane in which the elasticity is isotropic and can be characterized by five elastic constants.

Consider first the propagation of compressional waves. The elastic constants relevant for horizontal displacement (PH, SH wave) are given by

$$M_H = M_1 d_1 + M_2 d_2, \tag{B3-5-1}$$

and for vertical displacement (PV, SV waves),

$$M_V = \frac{1}{d_1 / M_1 + d_2 / M_2}. \tag{B3-5-2}$$

Therefore, M_H is always greater than $M_{\bar{v}}$ and, hence,

$$V_{SH} > V_{SV}, V_{PH} > V_{PV}. \tag{B3-5-3}$$

These are universal results that apply to *any layered structure* in which the wave length is much larger than the thickness of the layers. In long-wave-length deformation, the strain associated with PH (SH) motion must be homogeneous. Therefore, the shear strain in

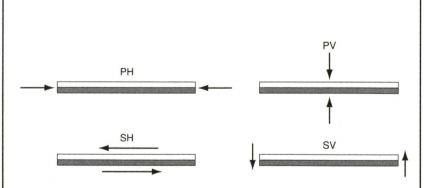

Fig. B3-5-1. Deformation of a layered material with two modes (one with horizontal displacement and another with vertical displacement).

each of the two layers must be the same for PH (SH) motion. Stress in each layer will then be different. In contrast, for PV (SV) motion, the strain in each layer is additive, but the stress must be the same. Therefore, in PH (SH) motion, a composite layered material behaves more stiffly than it does for a PV (SV) motion.

Thus, a comparison of P-wave anisotropy with S-wave anisotropy is one way to identify a layered structure. For other types of anisotropic structure, such a relation may not hold. If such a relation is not satisfied by the data, then one can conclude that a layered structure with a transverse isotropy is not responsible for the observed anisotropy.

The degree of anisotropy is given by

$$\frac{M_H - M_V}{M} \sim \phi(\frac{\Delta M}{M})^2, \tag{B3-5-4}$$

where M_H and M_V are the modulus for the horizontal and vertical deformation shown in the above diagrams and $\Delta M = M_1 - M_2$, $M = \sqrt{M_1 M_2}$, and ϕ is the volume fraction of the secondary phase (phase 1 or 2). Therefore the anisotropy of elastic wave velocity (V) is

$$\frac{\Delta V}{V} \sim \frac{\phi}{2}(\frac{\Delta M}{M})^2. \tag{B-3-5-5}$$

Note that the effects of layering are expressed by a second power of elastic-moduli contrast (therefore the nature of anisotropy does not depend on whether the second phase is softer or harder) and by a first power of volume fraction. Therefore, a large contrast in elastic constants is needed to cause significant shear wave anisotropy. A layered partially molten material is one possibility. In this case, a large anisotropy is expected for shear waves but not much of one for compressional waves. Another possible case is a layered structure with two equally sized phases, such as the case of the transition zone (Allégre and Turcotte 1986).

Horizontal shear is usually considered to result in a layered structure with *transverse isotropy*. Although this is correct in a crude sense, it is not necessarily true in a stricter sense. One important example is the melt-preferred orientation in a partially molten material. An experimental study on a partially molten peridotite by Zim-

(continued)

merman and his colleagues (1999) showed that the orientation of melt pockets is not parallel to the flow direction, but is tilted by ~ 20°. Therefore, the elastic anisotropy of such a sheared partially molten peridotite does not have transverse isotropy (see also Karato 1998).

1. There is no azimuthal anisotropy in the plane of layering (transverse isotropy).
2. For horizontally (vertically) layered structure, $V_{SH} > V_{SV}$ ($V_{SV} > V_{SH}$).
3. The strength of anisotropy is strongly dependent on the contrasts in elastic property between two phases. If there is 10% of the second material, for example, the elastic constants must differ by about 50% to create more than 1% anisotropy. If melt pockets are perfectly aligned in partially molten material, it would satisfy this condition.

How can this layering be related to flow? Though horizontal (vertical) layering corresponds to horizontal (vertical) flow to a good approximation, this relation is more complicated in reality. Consider the deformation of a material with two phases. These phases stretch as well as rotate by deformation. Theoretical studies have shown that the layered structure evolves slowly by the rotation of each phase and that the degree of rotation depends on the viscosity contrast of the two phases. Similarly, according to deformation experiments with partially molten materials, melt alignment does not exactly match with shear direction (Zimmerman et al. 1999; fig. 3-10). Note that azimuthal anisotropy also appears in this case.

The elastic anisotropy of a material with preferred orientation is determined by the elastic anisotropy of each mineral and the lattice preferred orientation of crystals. The elastic properties of constituent minerals for the mantle and the core have been investigated experimentally as well as theoretically, and much data have been accumulated for the anisotropy of these minerals. Most of these constituent materials are found to be anisotropic. It is especially notable that even cubic crystals such as ringwoodite (spinel) and (Mg,Fe)O have elastic anisotropy. Materials with cubic symmetry are not necessarily isotropic. MgO is a prominent example. As summarized in figure 3-11, MgO is strongly anisotropic, and anisotropy varies significantly with pressure. Many other minerals also show strong pressure dependence. This fact is important in the interpretation of the anisotropy of Earth's deep interior.

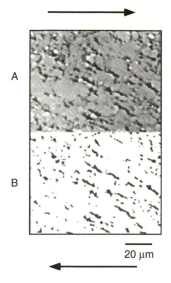

Fig. 3-10. When partially molten material (peridotite + 3% basaltic melt) is deformed, melt pockets (shown in black) align in a particular direction (after Zimmerman et al. 1999). The orientation of the melt pocket is nearly parallel to the shear direction, but there is a finite angle (~ 20°) between the melt orientation and the shear direction.

Once the elastic constants have been measured, we need to know the nature of the lattice preferred orientation, which is determined by a microscopic deformation mechanism (and recrystallization) as well as a macroscopic geometry of deformation. The microscopic mechanism of ductile deformation is briefly summarized in box 2-2. Recrystallization refers collectively to the formation of new crystal grains and the migration of grain boundaries, both of which often take place in deforming rocks. Recrystallization has a large influence on preferred orientation, but how it affects preferred orientation is not well known. In some cases, recrystallization only slightly modifies preferred orientation due to deformation, but in other cases it significantly changes the pattern of orientation. For simplicity I will briefly explain here only the case of preferred orientation due to deformation.

Microscopic processes of deformation affect the way in which lattice preferred orientation develops. Among several deformation mechanisms, lattice preferred orientation is produced by dislocation creep. It is not created by diffusion creep or grain-boundary migration. To understand this, consider how crystal orientation rotates by deformation (fig. 3-12). The

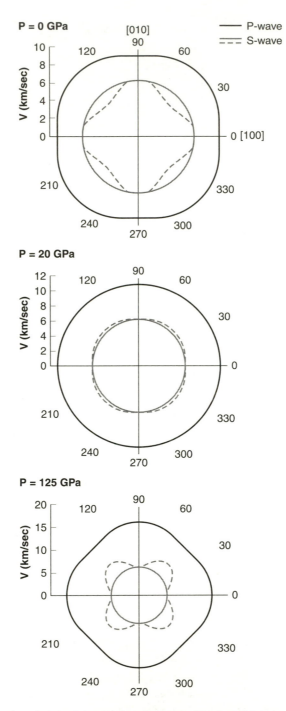

Fig. 3-11. Depth variation of the elastic anisotropy of MgO, which is one of major constituent minerals in the lower mantle. Note that anisotropy significantly varies with pressure (depth) (from Karato 1998).

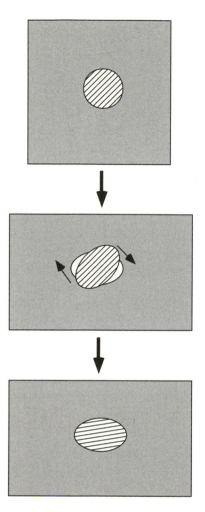

Fig. 3-12. The mechanism of crystal orientation due to deformation. A crystal in mineral aggregate deforms by some slip system (slip planes are shown as thin lines). This deformation does not necessarily match the deformation of the mineral aggregate. Crystal rotates to minimize this mismatch, resulting in the formation of preferred orientation.

shape of a crystal is modified by deformation. When deformation is realized by a dislocation motion, the shape of a crystal can change only in a certain manner. Thus, by rotating its orientation, a crystal tries to match the boundary condition imposed by its surrounding materials. This is the essence of the mechanism of the rotation of crystal orientation. Simply put, crystals tend to align to match the microscopic deformation by dislocation motion with the imposed macroscopic deformation geometry. Thus, to investigate anisotropic structure of lattice preferred orientation, it is important to study the slip systems of dislocations in minerals. When deformation is accommodated by diffusion creep, on the other hand, a crystal can make an arbitrary shape change, so its orientation does not rotate and no strong lattice preferred orientation will develop.

It has been long known in material science that lattice preferred orientation is formed by dislocation creep, not by diffusion creep or "superplasticity" (deformation due mostly to grain-boundary sliding). Recent experiments show that this holds also for minerals with a low degree of symmetry, such as perovskite (Karato, Zhang, and Wenk 1995a). Thus, we reach an important conclusion, that deformation does not always create lattice preferred orientation. In fact, many of the deformation mechanisms that result in lattice preferred orientation take place only at relatively high stresses. Consequently, whereas anisotropic structure develops in the boundary layers of convection where high stress is expected, anisotropic structure is unlikely in slowly deforming low-stress regions away from boundary layers. This idea can nicely explain seismic observations (fig. 3-8). The uppermost and lowermost parts of the mantle show strong anisotropy probably because they are the boundary layers in which the stress level is high. This idea has recently been tested by numerical simulations (McNamara, Karato, and van Keken 2001). The seismological model of Montagner and Kennett (1996) and Wookey, Kendall, and Barruol (2002) shown in figure 3-7 also shows some degree of anisotropy for the mid-mantle (500–800 km). This may indicate the presence of another boundary layer at these depths.

When deformation results in an anisotropic structure, its relation to deformation geometry as well as stress orientation is an important problem. In the 1970s, Adolfe Nicolas, in France, and his colleagues extensively investigated this problem for upper-mantle materials. Mainly by conducting the structural analysis of rocks that were actually deformed in Earth's upper mantle, they investigated the relation between deformation and anisotropy. Collaborating with Nick Christensen, in the United States, and others, Nicolas also measured the elastic anisotropy of these deformed rocks. By these efforts, the relation between deformation and seismic anisotropy was well established for most upper-mantle rocks by the mid-1970s (Nicolas and Christensen 1987). Among upper-mantle minerals, olivine has the strongest elastic anisotropy and is also easily subjected to plastic deformation. Thus, the anisotropic structure of the upper mantle is mostly determined by the lattice preferred orientation of olivine.

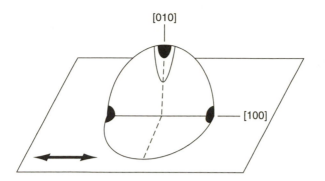

Fig. 3-13. The preferred orientation of olivine under water-poor conditions (after Nicolas and Christensen 1987). Arrows denote the direction of shear deformation. For this lattice preferred orientation, the compressional wave velocity is highest in the direction parallel to the shear direction and the polarization of the faster shear wave is parallel to the shear direction.

The observation of rocks deformed within the shallow portions of Earth often shows that the [100] axis and (010) plane of olivine tend to be parallel to the flow direction and the flow plane, respectively. This is consistent with the deformation mechanism of olivine with dislocation creep. In (anhydrous) olivine, [100] and (010) are the primary slip direction and the primary slip plane, respectively, and the microscopic slip direction and slip plane in each crystal tend to match the macroscopic flow (fig. 3-13). In this case, the faster propagation direction of compressional waves and the faster polarization direction of shear waves are almost parallel to the direction of mantle flow (fig. 3-13). For the horizontal flow, $V_{SH} > V_{SV}$, and for vertical flow, $V_{SV} > V_{SH}$. Using these relations, Tanimoto and Anderson (1984) estimated the pattern of mantle convection from the seismic anisotropy of the upper mantle (fig. 3-14).

Though upper-mantle anisotropy once seems to have been well understood, two problems have recently emerged. One is that the deeper portion of the upper mantle (> 200 km) turned out to have very weak anisotropy. If anisotropy is caused by olivine deformation, it is not obvious why anisotropy is reduced in the depths corresponding to the asthenosphere. The other problem is the regional variability of anisotropy. In particular, the patterns of anisotropy in the mantle wedge (the mantle above the subducting plates) and in the mantle beneath hotspot volcanoes are complicated. If one were to explain the complicated spatial distribution of anisotropy in the mantle wedge, assuming the traditional idea of

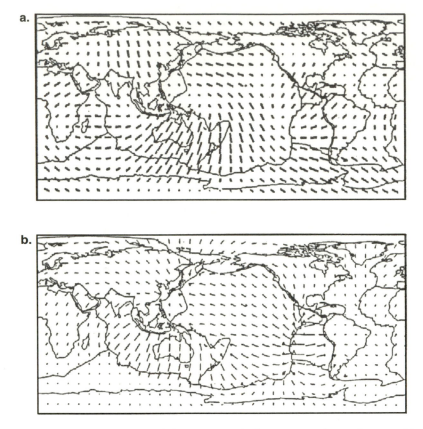

Fig. 3-14. Seismic anisotropy and the upper-mantle flow pattern (after Tanimoto and Anderson 1984). (a) Seismic anisotropy (the azimuthal anisotropy of the Rayleigh wave) of the upper mantle (around the depth of 200 km). The azimuthal anisotropy of the Rayleigh wave is similar to that of the compressional wave, so this diagram can be interpreted using fig. 3-13. The fast direction of the Rayleigh wave corresponds to the direction of mantle flow. (b) The flow pattern in the upper mantle is estimated from mantle convection models. To the first order, this is consistent with the above observation on anisotropy.

the relation between anisotropy and flow geometry, then one would need to invoke a highly complicated flow geometry there.

It appeared necessary to reexamine the formation mechanism of anisotropic structure from the microscopic point of view. The following two points have been established for the ductile deformation of olivine during the last ten years or so. First of all, under mantle conditions, either dislocation creep or diffusion creep is most likely the dominant deformation mechanism of olivine, and the conditions at which the dominant mechanism changes are close to the conditions expected in Earth's mantle (this is true for most other Earth materials). Experimental results indicate that diffusion creep is likely to dominate at high pressures, and this may be the reason for weak anisotropy in the deep upper mantle.

Second, the addition of water is found to change the lattice preferred orientation (Jung and Karato 2001). Although the effects of water in enhancing deformation of olivine (and other silicates) have been well known, the effects of water on lattice preferred orientation have not been well appreciated. Earlier studies on naturally deformed rocks have focused on relatively "depleted" rocks from the lithosphere, and most of the previous experimental studies were made under the conditions where the amount of water (hydrogen) dissolved in olivine was low. Given the experimental data by Mackwell, Kohlstedt, and Paterson (1985) and Yan (1992), which showed that the effects of water on dislocation motion in olivine are highly anisotropic, I proposed that the dominant slip system(s) in olivine might change at a high water content, leading to a change in lattice preferred orientation (Karato 1995). To test this hypothesis, we have conducted a number of laboratory experiments under high water content, and indeed found that the lattice preferred orientation in olivine changes as a function of water content (and stress) (fig. 3-15). Briefly, the dominant slip direction in olivine changes from [100] to [001] under high water content, and the dominant glide plane also appears to change with stress and/or water content.

Such changes in the lattice preferred orientation have a profound influence on the geodynamic interpretation of seismic anisotropy. For example, under high stress and high water content conditions, the direction of the faster S-wave polarization is perpendicular to the shear (flow) direction, as opposed to parallel to it in other conditions. It is possible that some of the seismological observations near convergence zones (subduction zones or collision zones) showing convergence-zone parallel shear-wave splitting may be due to the lattice preferred orientation of olivine under high water content and high stress conditions. In these regions, high water content and high stress are indeed very likely.

The V_{SH}/V_{SV} polarization anisotropy also changes with water content. When water content is low, horizontal (vertical) shear results in $V_{SH} > V_{SV}$ ($V_{SH} < V_{SV}$) anisotropy. However, when water content is high, $V_{SH} > V_{SV}$ ($V_{SH} < V_{SV}$) will result from vertical (horizontal) shear. This provides a

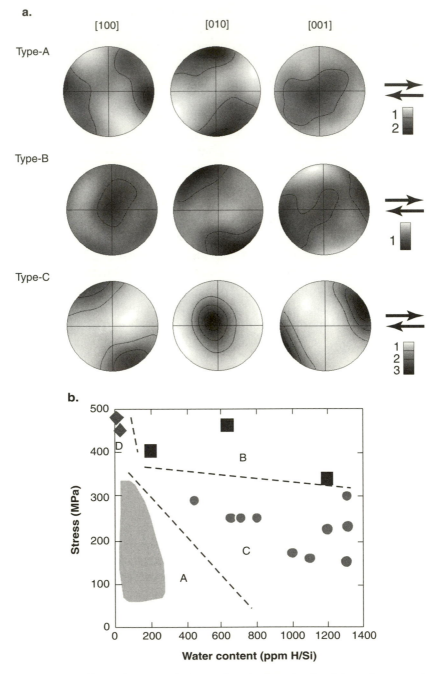

Fig. 3-15. The effects of water on lattice preferred orientation in olivine. When water is added, new types of lattice preferred orientation can be found. The pattern of lattice preferred orientation depends on the conditions of deformation. Part (a) denotes a typical lattice preferred orientation in olivine; (b) types of lattice preferred orientation at high temperatures ($T \sim 1,500-1,600$ K) as a function of water content and stress (after Jung and Karato 2001).

possible explanation for the observed anisotropy beneath hotspots such as Hawaii and Iceland (fig. 3-8). In these regions, upwelling hot materials likely contain high water content (there is geochemical evidence for high water content in the source regions of hotspot magmas). Therefore in the deep regions of upwelling current, anisotropy should be $V_{SH} > V_{SV}$. However, as hot materials rise, they will undergo partial melting, and much of the water would be lost (see chap. 2). At depths shallower than this, the upwelling materials will be dry (water-poor) and will show $V_{SH} < V_{SV}$. In other words, the depth of transition from $V_{SH} > V_{SV}$ to $V_{SV} > V_{SH}$ would correspond to the depth where substantial partial melting occurs and water (hydrogen) is removed from olivine. Finally when the hot materials reach the shallow asthenosphere, these dried materials will spread horizontally, leading to strong $V_{SH} > V_{SV}$ anisotropy (fig. 3-8).

A similar small-scale variation in anisotropy may occur in the continents. Although the major portion of the continental lithosphere is depleted with water (see chap. 2), petrological observations suggest a water-rich (or undepleted or enriched) region in the deep continental roots. Similarly, the continental upper mantle next to a collision zone will be enriched with water. Spatial variation in water content is likely to result in spatial variation in anisotropy.

The above relation between mantle flow and anisotropy is valid only for the olivine-rich upper mantle. Because it is based on the elastic and plastic properties of minerals, a completely different relation is expected for deep-mantle materials. Once this relation is known, the flow pattern of deep mantle can be inferred from seismic anisotropy, and considerable progress is expected for the study of mantle dynamics. The study of the deformation mechanism of deep-mantle materials is, however, still in its infancy. Unlike in the upper mantle, rocks directly sampled from deep mantle have not been discovered (some inclusions in diamond show evidence for deep mantle origin, but they are too small for the study of deformation microstructures), so we cannot apply the methods of structural geology. Furthermore, it is difficult to conduct deformation experiments at the high temperatures and pressures corresponding to deep mantle.

The situation is different for (Mg,Fe)O, one of the major constituents of Earth's lower mantle. This mineral is unique in that it has very large but highly depth dependent elastic anisotropy (fig. 3-11). Surprisingly, despite its simple crystal structure, the anisotropy of (Mg,Fe)O in the very deep portions of the lower mantle is among the largest of all major mantle minerals. Consequently, its lattice preferred orientation is of special importance for the geodynamic interpretation of seismic anisotropy. Daisuke Yamazaki and I conducted an experimental study of large-strain shear deformation of this material at moderate T/T_m (T_m is the melting temperature), similar to that in the lower mantle (Yamazaki and Karato 2002). The results showed that a strong lattice preferred orientation develops in (Mg,Fe)O by deformation by dislocation creep, and the resultant anisotropy is consistent with that found at the bottom of the lower mantle (the D″ layer): $V_{SH} > V_{SV}$ for horizontal shear. Therefore, a part of the observed anisotropy in the D″ layer can be explained by the lattice preferred orientation of (Mg,Fe)O (McNamara, van Keken, and Karato 2002). However, anisotropy in the shallow lower mantle cannot be attributed to (Mg,Fe)O because (Mg,Fe)O is elastically isotropic there.

New results on (Mg,Fe)O and a large data set of deformation mechanisms of many oxides and silicates provide a framework for the geodynamical explanation of anisotropy. The distinct spatial variation of anisotropy can be attributed to the spatial variation in stress levels in the convecting mantle. High stresses develop in or near (cold) boundary layers where deformation mechanism is likely to be dislocation creep. It is in these regions where strong anisotropy is observed (Karato 1998; McNamara, Karato, and van Keken 2001; McNamara, van Keken, and Karato 2002). Flow geometry in these regions can be inferred when the relation between flow geometry and anisotropy is known. Such is the case for olivine, although the relation is complicated due to the role of water (and stress). For most of other deep-mantle minerals, such relations are poorly known, except for (Mg,Fe)O.

In contrast to lattice preferred orientation, the importance of the layered structure is not well known. The most obvious origin of anisotropy is the anisotropy caused by aligned melt in the upper mantle as proposed by Aki (1968). However, evidence of aligned melt pockets is not found in the recent seismic experiment at mid-ocean ridge (Wolfe and Solomon 1998). It appears that the melt fraction in the upper mantle is either small or that a significant amount of melt is present only in very limited regions or that melt pockets are not aligned.

The origin of anisotropy in the D″ layer of the central Pacific remains uncertain. Numerical modeling shows that the stresses in these regions are low and deformation by dislocation creep is not likely. Aligned melt may be a cause for anisotropy, but it is not known why anisotropy due to aligned melt can be seen in the D″ layer but not at the mid-ocean ridges. Better understanding of structure of this region is critical to the better understanding of the origin of plumes.

3-4. SUMMARY AND OUTLOOK

Progress in seismology has been very rapid; due to improved digital seismological networks and rapidly increasing computer technology as well as theoretical developments in the data analysis. Several digital seismograph networks have been developed, including GEOSCOPE (France) and J-array (Japan). A new dense seismic network called USArray is planned in the United States, which will give us substantially higher-resolution images of Earth's interior. The situation in solid Earth science now is very much like that of astrophysics, where high-resolution images taken by new instruments such as the Hubble telescope have revolutionized our view of this universe.

Given these high-resolution images of Earth, we need to have a good *dictionary* to read them in terms of the structure and dynamics of Earth's interior. This dictionary must be written based on the *physics and chemistry of materials*. More precisely, we need to know how thermal and chemical convection affect the properties of materials that can be detected by seismological (and perhaps other) observations. Major progress has been made in this area, as I have summarized in this chapter. Improved high-pressure experimental techniques as well as theoretical calculations (such as quantum mechanical, first-principles calculations) have made important contributions. These studies have provided a working hypothesis for the cause of seismic anisotropy and the flow patterns in Earth's deep interior (fig. 3-16). However, two major issues remain challenging. First, the effects of chemical composition on elastic properties must be characterized in much more detail. Elastic properties are highly sensitive to pressure. Therefore, these studies must be made under high pressures (and temperatures). Second, as I have demonstrated in this chapter, seismic wave velocities are sensitive to plastic flow (through anelasticity and through texture [LPO] formation). However, the plastic flow properties under deep Earth conditions are poorly understood due to the difficulties in generating deviatoric stresses in a controlled manner and also to the difficulties in measuring deviatoric stresses under high confining pressures. However, major breakthroughs have been made in these two areas. Don Weidner, at the State University of New York at Stony Brook, has developed a powerful new technique of deviatoric stress measurements under high pressures using a synchrotron radiation facility (Weidner 1998). A new type of high-pressure deformation apparatus has been developed by Yamazaki and Karato (2001a), and another type of apparatus is being tested by Bill Durham, Don Weidner, and

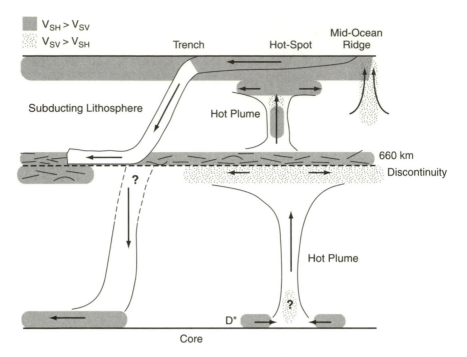

Fig. 3-16. The pattern of material circulation inferred from seismic anisotropy (modified from Karato 1998). Anisotropy is strong in convective boundary layers. In the top boundary layer, the lithosphere-asthenosphere, the flow pattern is estimated from anisotropy. Similar inference will be possible for the lowermost mantle (the D″ layer) as further progress is made in mineral physics. There are a few reports for anisotropy in the transition zone, which indicates partially layered mantle convection. V_{SH} and V_{SV} are horizontally and vertically polarized shear wave velocities, respectively.

Yanbin Wang and their colleagues. Combining results from these two areas (high-resolution seismology and high-pressure mineral and rock physics) will provide us with a much better understanding of this dynamically evolving planet.

FOUR • MANTLE CIRCULATION AND PROPERTIES OF EARTH MATERIALS

4-1. HISTORICAL BACKGROUND AND PROBLEM SETTING

Approximately 70% of the heat flux from Earth is released by plate tectonics to outer space. We now have a good understanding of how materials have been circulated by plate tectonics for the past ~ 200 million years, at least to the depth of about 700 km. At mid-ocean ridges, upwelling mantle materials partially melt, forming basaltic oceanic crust and residual "depleted" mantle (oceanic plate or lithosphere; see chap. 2). At ocean trenches, this oceanic lithosphere sinks into the deep mantle. This sinking process is also referred to as *subduction*. Subducting oceanic lithosphere has strongly negative buoyancy—in other words, it is heavier than the surrounding materials, which is the most important driving force for plate tectonics. Don Forsyth, then at Lamont Doherty Geological Observatory, and Seiya Uyeda, then at University of Tokyo, showed this through the analysis of the correlation between the velocities of plate motion and other parameters such as the length of mid-ocean ridges or the length of trenches associated with each plate (Forsyth and Uyeda 1975). The fate of subducted slabs can be traced by the hypocenters (locations of earthquakes) of deep earthquakes, but only down to ~ 700 km; below this depth, there is no seismicity (earthquake activity) at all (chap. 5). The question, then, is whether the subduction of the oceanic lithosphere is limited to this depth, and mantle convection is separated between the upper and lower mantle, or the subduction continues to the lower mantle, even though it cannot be observed by earthquake activities. There had been much debate about this issue since the advent of the theory of plate tectonics.

Much progress has been made to settle this issue during the past ten years or so. First, details of phase transformations of mantle minerals have been studied in the laboratory and the sequence of phase transformations and the nature of changes in densities have been characterized. The most

119

notable finding is that the ringwoodite to perovskite + magnesiowüstite transformation, thought to occur at the 660-km boundary, has a negative Clapeyron slope (chap. 1). In the 1990s, more realistic three-dimensional numerical simulations of mantle convection have been performed, using massive computers, in which the effects of negative Clapeyron slope are incorporated. These calculations showed surprisingly complicated interactions between convection and phase transformations. Some details of materials circulation are being mapped by high-resolution tomographic images. Such progress in the last decade or so is remarkable; our view of the dynamics of Earth's interior has undergone major modifications. In this chapter, I will review this progress and discuss some of the remaining important issues.

Mantle convection results in chemical differentiation. Earth is made of multicomponent materials, and when convection in Earth's mantle occurs, pressure-release partial melting occurs in hot upwellings to create melt and residual solids with different chemical compositions. At mid-ocean ridges, this partial melting produces mid-ocean ridge basalt (MORB) and the residual "depleted" peridotite (chap. 2). Similar chemical differentiation occurs associated with the formation of continental crust. The nature of chemical differentiation depends on the degree of partial melting, which in turn depends on the temperature and the chemical composition of the original material, such as the water content. Therefore, the chemical composition of mid-oceanic basalt is different from that of the continental crust. Based on petrological studies combined with seismological observations, the processes of chemical differentiation through partial melting in the upper mantle have been reasonably well understood (e.g., Ringwood 1975). Partial melting and the resultant chemical differentiation could have happened in the early Earth (during Earth's formation and/or the Archean period) at the greater depths.

Through studies of chemical compositions of volcanic rocks, geochemists have identified chemically distinct regions (sometimes called *reservoirs*) not only in the shallow portions, but also in Earth's deep interior (e.g., Hofmann 1997). Geochemical studies using radioactive isotopes also show that some of these regions have been separated from each other for more than one billion years. Therefore, it is clear that Earth's mantle contains several chemically distinct regions (reservoirs) that have maintained their identify over this period of time. However the exact locations, the origin of these reservoirs, and the mechanisms for their survival are poorly understood.

Two issues related to the origin and evolution of chemically distinct regions will be discussed in this chapter. The first is the fate of oceanic crust during subduction. The recycling of oceanic crust is one of the most important processes of geochemical cycling. The question is, where could it be separated from the rest of the mantle to form a large-scale geochemical reservoir? The second issue is the mechanisms by which geochemically distinct reservoirs have been separated from each other for 2 billion to 3 billion years. Physical properties, particularly the densities and rheological properties (viscosities) of chemically distinct reservoirs, have important influence on their behavior in convecting mantle.

Surprisingly, most of the previous studies on materials circulation in Earth ignored the role of rheological properties. In many cases, calculations were made assuming unrealistic rheological structures, such as homogeneous Newtonian fluid, ignoring the depth or material (lateral) dependence of rheological properties. One reason for this ignorance might be that we did not know much about the rheological properties of deep Earth materials. Although such a simplification is necessary and useful at a certain stage of research, the validity of some of the conclusions from such an idealistic model must be examined with caution. Results assuming highly idealistic materials properties would be applicable to a hypothetical planet made of such idealistic materials but might not be relevant to Earth. In this section, I will focus on the role of material properties, particularly rheology, on these problems.

4-2. THE INTERACTION BETWEEN THE SUBDUCTING OCEANIC LITHOSPHERE AND THE 660-KM BOUNDARY

A long-standing question regarding mantle convection is whether mantle convects as a whole (whole-mantle convection) or convects in a layered fashion. High-resolution tomographic studies in the late 1990s have provided almost definitive answer to this question: at least at the present time or in the recent past (within the last ~ 100 million years), whole-mantle scale circulation seems likely. The notion of slab penetration into the lower mantle was originally made by Tom Jordan and his students in the 1970s (Creager and Jordan 1974; Jordan 1977), and was confirmed twenty years later through higher-resolution imaging techniques by Steve Grand (1994) and Rob van der Hilst and his colleagues (van der Hilst et al. 1997). Observational data supporting this notion were already discussed in chapter 3. At the same time, however, it has also become clear that the pattern of convection is not simple whole-mantle circulation. Flattening of subducted slabs at the 660-km boundary and at the depths of ~ 900–1,200 km has been discovered beneath the western Pacific (van der Hilst et al. 1991; Fukao et al. 1992; Fukao, Widiyantoro, and Obayashi 2001). In contrast, a plate subducting beneath the South American continent seems to be penetrating into the lower mantle without notable deformation. The interaction between subducting plates and the mantle transition zone appears more complex than previously considered.

These seismic observations suggest that subducting slabs encounter some obstacles in the transition zone, and that the pattern of mantle convection is modified in various ways. In the transition zone, most of the mantle minerals undergo a series of phase transformations. Consequently, the complexities of the convection pattern in the transition zone are likely to be caused by the interaction of the convection pattern with phase transformations.

4-2-1. Phase Transformation and Convection

It has been known since the early 1970s that the 660-km boundary was an obstacle to mantle convection. Focal mechanisms of deep earthquakes in subducting slabs reaching 660 km show compressional stress in the direction of subduction, indicating a resistance force near the 660-km boundary (Isacks and Molnar 1971). A phase change with the largest density jump in the mantle (ringwoodite to perovskite + magnesiowüstite) occurs at the 660-km boundary, and this phase change has been considered to be the source of resistance.

In the early stage of mantle convection studies, the possibility of a viscosity increase due to a phase change was considered. It was implicitly assumed that a higher-density phase should have higher viscosity. Also, the results of earlier analysis of the shape of Earth suggested higher viscosity in the deep mantle. In the late 1970s, however, a number of papers claimed that viscosity did not significantly increase with depth, and the interest in this viscosity issue drifted away. As briefly discussed in chapter 1, this argument for nearly depth-independent viscosity had no sound physical basis. Geophysical analyses suggesting high viscosity in deep mantle have appeared since the mid-1980s, and many scientists now believe that the deep mantle is significantly more viscous than the upper mantle (asthenosphere). At any rate, since the viscosity of Earth's mantle was not well understood in the past, scientists' attention moved to other, better-constrained physical properties such as density.

The majority of studies on the role of phase transformation on mantle convection have focused on the effects of the changes in density and entropy. Most of the phase transformations in Earth's mantle are first-order phase transformations in which both the density and the entropy of materials change discontinuously (density and entropy are the first derivatives of free energy, and, therefore, those transformations in which discontinuous changes in density and entropy occur are called first-order transformations). The densities of high-pressure phases are always greater than those of low-pressure phases. However, changes in entropy depend on particular phase transformations. The olivine → wadsleyite and wadsleyite → ringwoodite transformations are exothermic transformations in which high-pressure phases have lower entropy. In contrast, the ringwoodite → perovskite + magnesiowüstite transformation is an endothermic transformation in which the high-pressure phases have higher entropy than the low-pressure phase. This is due to a drastic change in crystal structure associated with a transformation to perovskite in which a silicon atom is surrounded by six oxygen atoms rather than four (see chap. 1).

The effects of entropy change on convection are different from those of density change. Consider an endothermic transformation such as the ringwoodite → perovskite + magnesiowüstite transformation in a descending slab. Because this is an endothermic transformation, some heat is absorbed upon the phase transformation. Therefore, descending materials will be cooled, extra negative buoyancy is obtained, and convection is enhanced. The effect of density change is the opposite. Along the downgoing cold current, the endothermic phase transformation would occur at a greater depth than it would in the surrounding hot regions. Consequently, there is additional buoyancy in the descending current, and, hence, the convection is prevented by the density change (fig. 4-1). The effect of density will be more important when the amplitude of temperature anomalies associated with convection is large. In contrast, the magnitude of latent heat release does not depend strongly on the velocity of convection. Consequently, the effect of density change will dominate when the convection velocity is high (i.e., convection at large Rayleigh numbers).

The Rayleigh number is a parameter that determines if thermal convection takes place or not: convection takes place for a Rayleigh number larger than $\sim 10^3$. The Rayleigh number of the whole-mantle system is around 10^5–10^7, and convection must take place in Earth's mantle. The strength and scale of convection also depend on the Rayleigh number. For higher Rayleigh numbers, convective flow becomes more vigorous and is characterized by a number of smaller-scale cells. The velocity of a plate thus increases with the Rayleigh number, and this results in the decrease

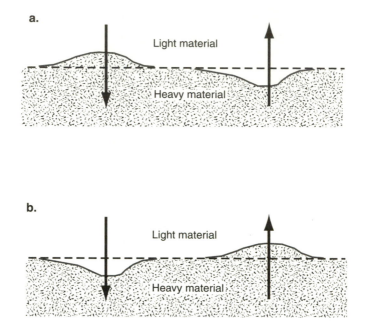

Fig. 4-1. The interaction between the phase boundary and mantle convection. For the high Rayleigh numbers expected for the Earth's mantle, the deformation of the phase boundary has a large effect on convection. (a) The case of a positive Clapeyron slope. In a region where cold material (a plate) is sinking, phase transformation takes place at a shallower level around the cold material. As a result, negative buoyancy increases, giving an extra driving force for convection. A similar effect to enhance convection occurs where hot material (plume) is upwelling. (b) The case of a negative Clapeyron slope. In a region where cold material (plate) is sinking, phase transformation takes place at a deeper level around the cold material. As a result, negative buoyancy decreases, generating a resisting force for convection. A similar effect occurs where hot material (plume) is upwelling.

of slab temperature as the Rayleigh number increases. At the same time, the width of down-going or upwelling current becomes thinner with an increasing Rayleigh number. For the high-Rayleigh-number convection expected for Earth's mantle, therefore, the effect of a deformed phase boundary is more significant than that of latent heat, and the 660-km phase boundary acts as a barrier for convection. This was confirmed by the numerical studies of mantle convection.

The next question is whether this 660-km boundary can be a strong enough barrier to promote layered convection. Ulrich Christensen, then at the Max Planck Institution in Germany, and Dave Yuen, at the University of Minnesota, studied this problem in great detail (Christensen and Yuen 1984, 1985). They considered not only the density change due to phase transformation, but also the density change due to chemical composition. As explained in chapter 1, the upper and lower mantle may have different chemical compositions, and if this is the case, the density difference corresponding to the compositional difference can have a great influence on the pattern of convection. The results of their study are summarized in figure 4-2, in which the style of convection is classified as a function of the magnitude of the Clapeyron slope and the density difference of the chemical origin. The degree to which the distortion of the phase boundary affects convection is determined by the competition between the buoyancy forces caused by thermal expansion and the modification of the buoyancy forces due to the distortion of the phase boundary. Therefore, the change in the style of convection due to the effect of phase transformation is characterized by a nondimensional parameter,

$$P_b \equiv \frac{\Gamma\left(\frac{\Delta\rho}{\rho}\right)}{\alpha\rho g h},$$

(4-1)

where Γ is the Clapeyron slope, $\left(\frac{dP}{dT}\right)_{eq}$; $\Delta\rho$ is the density jump associated with the phase transformation; α is the thermal expansion; and h is the

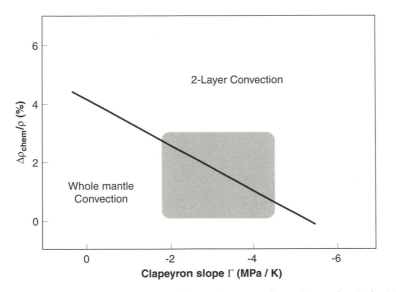

Fig. 4-2. The effect of density difference due to phase transformation or chemical origin on the pattern of convection (after Christensen and Yuen 1984). Shaded region corresponds to estimated Earth conditions.

thickness of the convecting layer. This parameter describes the ratio of the density change due to phase transformation and the density change due to thermal expansion. For an h of 3,000 km, $P_h \approx -0.2$, and for an h of 600 km, $P_h \approx -1$ ($\Gamma = -2.8$ MPa/K and $\alpha = 2 \times 10^{-5}$ [1/K]). Buoyancy due to phase transformation is thus comparable to the buoyancy due to thermal expansion. The pattern of convection becomes layered when the absolute value of P_h exceeds some critical value.

Christensen and Yuen concluded that the effect of the 660-km boundary on convection is relatively small under realistic conditions (Clapeyron slope ≈ -2 to -4 MPa/K, and a chemical density difference of 0 to 2%). Their calculation was, however, conducted in two dimensions, and the effect of the phase boundary was found to be larger in the subsequent three-dimensional calculations. I will explain this shortly, and before doing that, I would like to mention one more important result in the study of Christensen and Yuen (1985): the dependence of this effect on the Rayleigh number. It was found that the tendency for layered convection is enhanced as the Rayleigh number increases. According to their study, the critical phase transition buoyancy parameter for layered convection P_{hc} is related to the Rayleigh number as $P_{hc} \sim -4.4 Ra^{-0.2}$. As the Rayleigh number increases, therefore, the condition for layered convection is relaxed, and layered convection becomes more likely to take place. A higher Rayleigh number leads to a larger temperature difference between a boundary layer (such as a plate) and its surroundings, and the effect of phase transition buoyancy thus becomes more significant. From this, we can derive an important conclusion that as Earth gradually cools off, the pattern of mantle convection may have evolved from layered convection to whole-mantle convection (S. Honda 1995).

Christensen and Yuen also studied the effect of the convective pattern on heat (fig.4-3). As the parameter P_h varies, the pattern of convection changes as well as the efficiency of heat transport (which is shown by the Nusselt number [box 4-1]). When the pattern of convection changed from layered to whole-mantle, therefore, the cooling rate of the Earth must have rapidly increased. As a possible result, the cooling rate of the core may also have increased, and the emergence of the inner core may be related to the pattern change of mantle convection. Breuer and Spohn, in Germany, discussed this idea with reference to the data on the changes in the strength of the magnetic field over time and suggested that the Archean to Proterozoic transition might have been caused by the change in style of mantle convection (Breuer and Spohn 1995).

Such an idea of transition in the style of mantle convection with time is

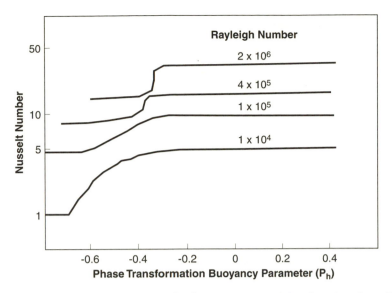

Fig. 4-3. The relation between convective heat transport and the phase transformation buoyancy parameter for the 660-km phase transformation (after Christensen and Yuen 1985). The Nusselt number denotes the efficiency of heat transport (box 4-1). When the phase transformation buoyancy parameter is negative and its absolute value is large, convection becomes layered and the efficiency of heat transport decreases. This is because a thermal boundary layer develops between the two layers, where heat is transferred by inefficient thermal conduction.

speculative because there is no way to directly test this hypothesis. However, this hypothesis is worth detailed investigation for the following reasons. First, such a transition is physically plausible. Second, as will be discussed soon, the present-day mantle appears to show features that the conditions there are on the margin between whole-mantle convection and layered convection. If so, in the past, where the Rayleigh number is likely to have been higher than it is today, layered convection would have occurred. Third, as Claude Allègre, at Institut de Physique du Globe in Paris, discussed (1997), such a change in convection style provides a possible explanation for the geochemical observation showing the persistent presence of two or more distinct chemical reservoirs for billions of years (this point will be discussed later, in sec. 4-3).

In the 1990s, more powerful computers became available, and this type of calculation began to be applied to more realistic models. Satoru Honda, together with Dave Yuen's group at the University of Minnesota and Paul Tackley's group at Caltech, conducted, almost at the same time, the three-

Box 4-1. The Nusselt Number

Convection carries heat away from a viscous layer more effectively than thermal conduction does. The efficiency of heat transfer by convection is measured by a nondimensional parameter called the *Nusselt number, Nu,* defined as

$$Nu \equiv \frac{[\textit{heat carried by convection}]}{[\textit{heat carried by thermal conduction}]}. \qquad (B4\text{-}1\text{-}1)$$

When a vigorous thermal convection occurs, then thermal boundary layers are formed at the top and the bottom of a fluid layer. The Nusselt number is related to the thickness of thermal boundary layer as $Nu \sim L/h$, where h is the thickness of thermal boundary layer and L is the thickness of a fluid layer. Consequently, a layered convection, which involves a larger number of thermal boundary layers than a whole-mantle convection (fig. 1-9), results in lower efficiency in heat transfer.

dimensional calculation of mantle convection with phase transformations. In a parallel study, David Bercovici and others made a theoretical analysis of this instability using an analytical model (Bercovici, Schubert, and Tackley 1993). They showed that the effect of phase transformation is larger in three dimensions than in two dimensions; that convection becomes temporarily stagnant at the 660-km boundary; and that after cold and dense materials accumulate above the 660-km boundary, they suddenly sink into the lower mantle (Bercovici, Schubert, and Tackley 1993; S. Honda 1993; Tackley et al. 1993). Dave Yuen calls this a "flushing event," and Paul Tackley, a "mantle avalanche." Whatever it might be called, this intermittently layered convection, or hybrid mode of convection, is consistent with seismic tomography, and this style of convection is widely believed to be close to the actual situation in the present-day Earth.

A closer look, however, reveals a number of differences between this type of numerical simulation and the real Earth. In the three-dimensional numerical simulation, platelike sinking materials stagnate at the phase boundary. By accumulating and forming a cylindrical form, they can finally sink into the lower mantle (this tendency can be more clearly seen in calculations with a realistic spherical geometry). According to this calculation, therefore, downwelling in the lower mantle must have a plume (cylindrical) shape. Shigenori Maruyama, at Tokyo Institute of Technology, calls it a cold plume (1994).

This convection model, however, is not quite consistent with the results of seismic tomography. High-velocity anomalies in the lower mantle beneath the American and southern Asian continents, which are often taken as the evidence for whole-mantle convection, are not cylindrical but planar (van der Hilst, Widiyantoro, and Engdahl 1997). Furthermore, the deformation style of a subducting slab in the mantle transition zone varies from place to place. What accounts for these discrepancies from the model predictions? I think that a key to solving these problems is the rheology of subducting slabs. Though the effect of rheology has been largely ignored in the past, it plays an important role in the deformation of cold materials such as subducting slabs, as explained in the next section.

4-2-2. The Rheological Effects on Convecting Patterns

In most convection studies, the effect of plate rheology has not been given detailed consideration. If subducting slabs are very stiff, one might intuitively think, then however strong a barrier the 660-km phase boundary is, slabs would penetrate into the lower mantle, resulting in whole-mantle convection. On the other hand, if slabs are soft, they may be deformed as they encounter a barrier, and the convection pattern may be approximately layered. Such a point was first studied by Christensen and Yuen (1984) and was later investigated in more detail by Geoff Davies (1995). Using the results of simulations incorporating the high strength of subducting slabs, Davies argued that the role of the 660-km discontinuity in a real Earth would be less than seen in numerical modeling by S. Honda and his colleagues (1993) and Tackley and his group (1993), where the rheological contrast between slabs and the surrounding materials was ignored. Thus, the strength of subducting slabs is an important issue. How strong are they?

A subducting slab is cold, and thus dense, and therefore sinks into the deep mantle. Cold material generally has a high viscosity, so one may naturally expect that slabs have high viscosity. This expectation is, however, inconsistent with recent findings from seismic tomography. A slab is colder than its surrounding mantle; of course the temperatures in a slab vary from place to place, but on average temperatures in a slab are lower than those in their surroundings by several hundred degrees. Using the well-known temperature dependence of mantle viscosity, then, we should expect subducting slabs to be more viscous than the ambient mantle by several orders of magnitude. Such high-viscosity slabs would not deform very much, even for the geological time. Recent results of high-resolution tomography, however, showed significant deformation of slabs in the mantle transition zone (500–700 km) beneath the western Pacific (chap. 3). Actual slabs seem to have much lower viscosity than expected from their low temperatures.

Soon after this significant deflection of slabs at around the 660-km discontinuity was found, Rob van der Hilst and Tetsuzo Seno proposed a model to explain it (1993). They noted that there is some correlation between slab deformation (at around the 660-km discontinuity) and the velocity of the migration of a trench. Trenches are usually fixed (with respect to the deep mantle), but in certain areas they move. When trench migration occurs and a subducted slab encounters large resistance forces at the 660-km discontinuity, then slabs will lie on top of the 660-km discontinuity (fig. 4-4). Laboratory experiments using a syrup in a tank also show this behavior (Guillou-Frottier, Buttles, and Olson 1996). This is one possible explanation for slab deflection at the 660-km discontinuity.

However, the correlation between trench migration and slab deformation is not so obvious. Although significant trench migration occurs in South America, there is no marked deformation of slabs near the 660-km discontinuity. By looking more closely at the results of seismic tomogra-

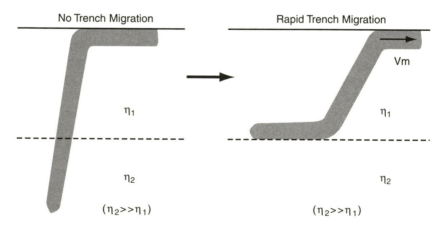

Fig. 4-4. A schematic diagram showing the effects of trench migration for slab deformation (after van der Hilst and Seno 1993). When a slab encounters some resistance at a boundary (say, the 660-km discontinuity), and trench migration occurs, a slab will lie on top of the boundary, if the slab is weak enough.

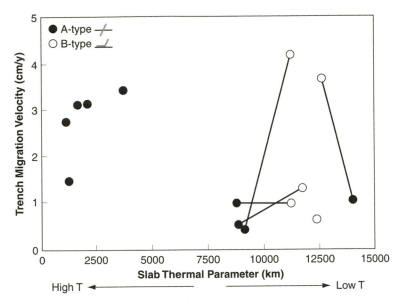

Fig. 4-5. A classification of the style of deformation of subducted slabs based on seismic tomography (after Karato et al. 2001). The morphology of slabs is classified into two categories and plotted against the slab thermal parameter, which characterizes slab temperature (see box 4-2). A good correlation is noted between the slab thermal parameter and the style of slab deformation. A cold, rapidly subducting plate is significantly deforming. The effects of trench migration velocity seem of secondary importance.

phy, we see that slab deformation is better correlated with the slab thermal parameter (box 4-2), namely, slab temperature. Slabs in the western Pacific that have higher slab thermal parameters (colder slabs) appear to deform more significantly than the slabs in the eastern Pacific (warmer slabs) (fig. 4-5). This is a rather odd observation, given the expected temperature of subducted slabs; the slabs subducting at the eastern Pacific are young, and thus hot, whereas the slabs subducting in the western Pacific are old and should be cold. Colder material must have higher viscosity and must be less likely to deform. This well-known temperature dependence of the rheology of materials seems to contradict the way slabs actually deform in the deep mantle as revealed by seismic tomography. How can we solve this paradox?

To resolve this problem, we need to investigate the rheology of subducted slabs in some detail. The first important contribution to the mechanics of the deformation of the lithosphere was the work by Chris

Box 4-2. Slab Thermal Parameter

The temperature profile of a descending (subducting) slab is controlled by the initial temperature, heat generation, and thermal conduction. When heat generation due to radiogenic elements, chemical reactions, and viscous heating is neglected, the temperature profile is controlled mainly by the initial temperature and the rate of thermal diffusion. The initial temperature at the time of subduction at the trench is controlled by the age of the plate at the trench (t_{sub}). The initial temperature at the time of subduction may be characterized by the depth (z) to a certain temperature as (see box 2-1)

$$z \sim \sqrt{\kappa t_{sub}}, \qquad (B4\text{-}2\text{-}1)$$

where κ is thermal diffusivity. As the lithosphere subducts, it will be warmed by thermal conduction from above. Consequently, although the initial temperature profile is monotonic with depth (warmer in the deeper portions), the slab thermal structure will

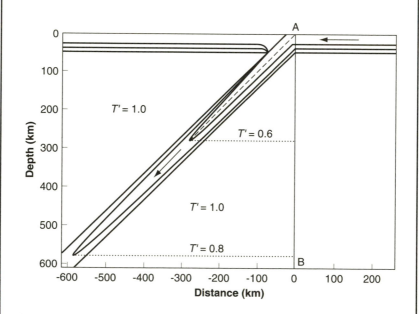

Fig. B4-2-1. Temperature distribution in a subducting slab (McKenzie, 1969). T' is the background temperature. The length of a thick dashed line is proportional to the slab thermal parameter.

be different. The temperature at the two surfaces of a slab is high because they are in contact with hot ambient mantle, and the central portion will be cooler. As a result, a cold tongue will develop from the shallow regions to a certain depth (fig. B4-2-1). A cold region could penetrate to a depth at which the time for warming ($t_{th} \sim L/v_{sub}$, where L is the length of a cold tongue) is equal to the time for cooling at the ocean floor (t_{sub}). Consequently,

$$t_{warm} \sim \frac{L}{v_{sub}} = t_{sub};$$ (B4-2-2)

hence,

$$L \sim v_{sub}t_{sub} \equiv \Phi,$$ (B4-2-3)

where Φ is the slab thermal parameter. Therefore, the slab thermal parameter (which has a dimension of length) represents a depth to which a given temperature profile penetrates into the deep mantle. The larger this parameter is, the cooler the slab.

Goetze and Brian Evans, at MIT (1979). They showed that when modeling deformation of relatively cold lithosphere, the widely used creep law is not adequate. Rather a "yield stress" type rheology must be used, which will limit the stress to a low value (box 4-3).

However, in order to understand the interaction of slabs and the 660-km discontinuity, we need to extend their work to deep Earth conditions. To this end, we need (1) to conduct deformation experiments under the pressure and temperature conditions corresponding to the mantle transition zone and (2) to estimate the role of phase transformations on rheological properties. To conduct deformation experiments under high pressures and high temperatures, we have developed a new method, which allows us to do semiquantitative experiments under pressures up to ~ 20 GPa and temperatures up to ~ 2,000 K (Karato and Rubie 1997; Karato, Dupas-Bruzek, and Rubie 1998). This method significantly extends the previously available range of pressure and temperature for experimental techniques (< 3 GPa, < 1,600 K). Using this method, the first deformation experiment for ringwoodite (silicate spinel), which is one of important minerals in the mantle transition zone, was conducted (Karato, Dupas-Bruzek, and Rubie 1998; fig. 4-6). The result of this experiment shows

Box 4-3. The Peierls Stress, Yield Stress and the Rheology of the Bending Lithosphere

Deformation of the lithosphere is a critical process for plate tectonics. It is important for the understanding of the mechanics of subduction at trenches as well as of the deformation of subducted slabs in the deep mantle. Temperatures in the lithosphere are low compared to other regions of Earth. If one were to use a conventional rheological flow law such as (box 2-2),

$$\dot{\varepsilon} = A\sigma^n exp[-\frac{H^*}{RT}], \tag{B4-3-1}$$

one would get an enormous stress (or effective viscosity, $\sigma/\dot{\varepsilon}$), and a bending would appear impossible. A paper by Goetze and Evans

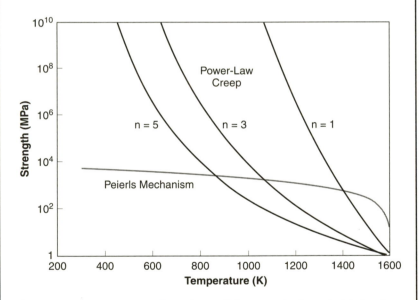

Fig. B4-3-1. The temperature dependence of the strength of a material at a given strain rate. For a power-law creep rheology with a small n, the strength of a material becomes very high at low temperatures. However, if n is large, the strength does not increase with decreasing temperature as much as for the cases with a smaller n. An exponential flow-law such as the flow law for the Peierls mechanism corresponds to $n = \infty$, and the strength is moderate even at low temperatures. (The strength is normalized such that strength = 1 MPa at 1,600 K.)

(1979) made an important contribution toward solving this problem. They showed, based on the experimental results, that a more appropriate flow law at high stress and low temperature is

$$\dot{\varepsilon} = A'\sigma^2 exp\left\{-\frac{H^*\left(1-\frac{\sigma}{\sigma_P}\right)^2}{RT}\right\},$$ (B4-3-2)

where σ_P is the Peierls stress. The difference between these two flow laws is shown schematically in figure B4-3-1. The basic point is that with the flow law illustrated in equation (B4-3-2), the flow stress does not go to infinite at $T \to 0$, although it would with the more conventional flow law, equation (B4-3-1). Rather the flow stress at $T \to 0$ for equation (B4-3-2) is $\sigma \to \sigma_P$. Physically, the Peierls stress is the stress needed to move crystal dislocations at T = 0 K, and is equivalent to "yield stress." In other words, in any real crystal plastic deformation is possible even at very low temperatures at a reasonable stress level because dislocations can go over the potential barrier (see box 1-5) if a sufficiently high stress is applied. This concept was first quantitatively formulated by a German physicist Rudolf Peierls, then at the University of Birmingham, England. The stress needed to move dislocations at $T = 0$ K is referred to as the Peierls stress.

that after applying the appropriate scaling law (box 4-4), when the grain size of ringwoodite is less than about 0.1–1 mm, its viscosity decreases in proportion to its grain size. A typical grain size of minerals in the oceanic lithosphere is ~ 2–3 mm. However, after a phase transformation, grain size can be reduced to an order of 1 micron. Under such conditions, the viscosity of materials can be reduced by a factor of 10^6 or larger. Therefore, the next issue is to understand how grain size is determined during a phase transformation in Earth.

While conducting deformation experiments under ultrahigh pressures, I also collaborated with Mike Riedel in Germany to theoretically investigate the scaling law of grain-size change upon a phase transformation. Grain-size reduction by a phase transformation has been often found in laboratory experiments, and it was often argued that a phase transformation could promote grain-size-sensitive creep and hence result in rheological weakening (Vaughan and Coe 1981; Ito and Sato 1990). How-

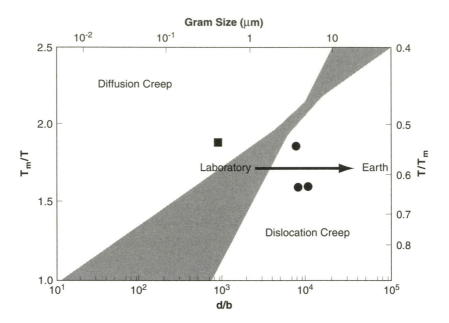

Fig. 4-6. A deformation mechanism map of ringwoodite (silicate spinel) (after Karato et al. 1998). When the grain size is small, deformation is achieved by diffusion creep, which is sensitive to grain size, so viscosity decreases with the decrease of the grain size. At the high stress (or high strain rate) used in experiments, this grain-size dependency is observed for grain sizes less than 1 − 2 μm. When scaled to mantle conditions (i.e., much lower stress and a slower strain rate (see box 4-4)), grain-size dependency is found to be important for grain sizes less than 0.1 − 1 mm. T_m, melting point; d, grain size; and b, length of Burgers vector.

ever, the observation of a small grain size in a laboratory experiment does not necessarily mean that we expect a similarly small grain size in Earth. Time scales are very different between laboratory experiments of phase transformations and phase transformations in Earth, and the grain size expected in Earth might be very different. To clarify this issue, we examined the microscopic physics of phase transformation and established a method for estimating the degree of grain-size reduction on the geological time scale from the result of laboratory experiments (Riedel and Karato 1997) (box 4-5). The basic concept of this model has been well known in materials science and is used in synthesizing diamonds, for example; grain size is determined by the ratio of the nucleation rate to the growth rate of a new phase. Frequent nucleation with a slow growth rate

Box 4-4. Scaling Laws in Rheological Properties

One of the difficulties in the laboratory study of rheology is the fact that the rate of deformation in laboratory studies is necessarily much faster than the rate of deformation in Earth. Therefore, any application of laboratory results to geological problems involves some sort of extrapolation. Let us discuss this problem in relation to grain-size dependence. In most cases, polycrystalline materials (rocks, for example) show two types of deformation behavior at relatively high temperatures. At high stress and with a coarse grain size, they deform by dislocation creep, whereas at low stress and with a small grain size, they deform by diffusion creep. When temperature and pressure conditions in the experimental study are similar to those of Earth, stresses used in the laboratory will be much higher than those in Earth.

The flow laws in each regimes are given by (see box 2-2),

$$\dot{\varepsilon} = A\sigma^n \qquad\qquad (\text{B4-3-1})$$

for dislocation creep and

$$\dot{\varepsilon} = B\frac{\sigma}{d^m} \qquad\qquad (\text{B4-3-2})$$

for diffusion creep.

Therefore the transition condition from dislocation to diffusion creep is given by

$$d = \left(\frac{B}{A}\right)^{\frac{1}{m}} \sigma^{-\frac{n-1}{m}} = \left(\frac{B}{A^{\frac{1}{n}}}\right)^{\frac{1}{m}} \dot{\varepsilon}^{-\frac{n-1}{nm}}. \qquad\qquad (\text{B4-3-3})$$

Thus,

$$\frac{d(\dot{\varepsilon}_1)}{d(\dot{\varepsilon}_2)} = \left(\frac{\dot{\varepsilon}_1}{\dot{\varepsilon}_2}\right)^{-\frac{n-1}{nm}}. \qquad\qquad (\text{B4-3-4})$$

A relation such as (B4-3-4) is referred to as a scaling law, which is needed to extrapolate laboratory results obtained at laboratory strain rates to geological strain rates. The scaling law (B4-3-4) shows that the transition from dislocation to diffusion creep in the
(*continued*)

laboratory occurs at a much smaller grain size than in Earth. In other words, in a laboratory study, we use much smaller grain size than in Earth and determine the conditions at which transition from dislocation creep to diffusion creep occurs. Then using the flow laws in each mechanism, we extrapolate these results to smaller strain rates in Earth. If we use a sample whose grain size is similar to that in Earth, we would not be able to detect a change in deformation mechanism in the laboratory. To conduct geologically relevant deformation experiments, we need to use samples whose grain size is much smaller than that in Earth.

leads to a small grain size (therefore, a trick to make a large diamond is to minimize the nucleation rate, keeping the growth rate large). Consequently, strongly nonequilibrium phase transformation at lower temperatures would lead to a small grain size. This theory therefore predicts that the degree of grain-size reduction associated with a phase transformation is sensitive to temperature. Although grain size decreases for phase transformation at low temperature, it rarely changes at high temperature (fig. 4-7). This result is also supported by experiments. In addition, the growth of grains after the completion of a phase transformation is also sensitive to temperature. The growth rate is lower (higher) at lower (higher) temperatures. Therefore, for two reasons, the grain size in a subducting slab after the phase transformation will be smaller in colder slabs: because (1) the size of grains at the completion of a phase transformation is smaller for lower temperatures and (2) the subsequent grain growth will be slower at lower temperatures. Consequently, the decrease in viscosity by grain-size reduction is important in cold slabs, but not in warm slabs. Karato, Riedel, and Yuen (2001) quantified these points through theoretical analysis.

Using these results on grain-size effects, Karato, Riedel, and Yuen (2001) investigated the rheological structure of subducted slabs. For simplicity, they ignored the role of a thin paleo-oceanic crust, and focused on the rheology of the main part of slabs (the rheological structure of paleo-oceanic crust will be discussed in the next section). Because the rheological properties depend on the stress (and other parameters), we need to know the distribution of stress to determine the rheological properties. In this calculation, we assumed that the deformation of slabs occurs as a bending corresponding to a certain bending moment, which is, we as-

Box 4-5. Grain-Size and the Kinetics of Phase Transformation

The size of grains often changes after a phase transformation. Imagine you are making ice from water. If you cool water rapidly, you will end up with fine-grained ice. In contrast, when you cool it very slowly, you will get a coarse-grained ice.

The basic physics behind this can be understood by considering the processes of phase transformation into two parts (*nucleation* and *growth*) and their mutual interaction. In order to make a new phase from an old phase, one needs to nucleate new phases, often at the grain boundaries of the old phase. Once nuclei of new grains are formed, they grow and impinge on each other to complete the phase transformation (fig. B4-5-1). These processes occur in a statistical fashion.

The rate of nucleation (on grain boundaries), I, has a dimension of $(1/m^2 s)$. The rate of growth, Y, has a dimension of (m/s). Therefore from these parameters, we can find two parameters that have dimensions of time and length,

$$\tau \equiv \left(IY^2\right)^{-1/3} \qquad\qquad\text{(B4-5-1)}$$

and

$$\delta \equiv \left(\frac{Y}{I}\right)^{1/3}, \qquad\qquad\text{(B4-5-2)}$$

respectively. τ is often called the Avrami time, and δ is the Avrami length. Numerical modeling shows that the time needed to complete

(*continued*)

a phase transformation is approximately given by τ, and the size of grains after a transformation is approximately given by δ (Riedel and Karato 1997). Equation (B4-5-2) shows that when the nucleation rate is high (as in the case of the fast cooling of water) and growth rate is slow, then the size of the grains will be small. In contrast, when the nucleation rate is low but the growth rate is high, then one gets a large grain size. It is also useful to use another relationship. From equations (B4-5-1) and (B4-5-2), one can easily show that

$$\delta = Y\tau. \tag{B4-5-3}$$

Now, both nucleation and growth rates are dependent on temperature (and pressure). In a real geological situation, such as the phase transformation in subducting slabs, the rate of nucleation is roughly determined by the rate of subduction and is only weakly dependent on temperature. When temperature is very low, then effective transformation begins after the depth of materials becomes much greater than the depth at which equilibrium transformation would occur. In this case, the rate of transformation is not slow, despite low temperatures, because the driving force for the transformation is large. Similarly, when the temperature is high, then transformation will occur immediately after a material goes deeper than the equilibrium depth, so the rate of transformation will not be so fast, despite high temperature, because the driving force is small. Therefore, for phase transformations in subducting slabs, one can approximate that τ is nearly independent of temperature. Under these circumstances, the temperature dependence of the size of the grains after a transformation is largely controlled by the temperature dependence of the growth rate,

$$\delta \propto Y \propto exp\left[-\frac{H_g^*}{RT}\right]. \tag{B4-5-4}$$

Therefore, the size of grains after a transformation is strongly dependent on the temperature at which the transformation occurs: the lower the temperature, the smaller the grain size. By combining this result with the flow law for diffusion creep in equation (B2-2-3), one gets

$$\dot{\varepsilon} \propto exp\left[-\frac{mH_g^* - H_n^*}{RT}\right]. \tag{B4-5-5}$$

For a reasonable range of parameters in this equation, one has mH_g^* $- H_\eta^* < 0$. Consequently, when grainsize is controlled by the kinetics of transformation and is small enough to cause diffusion creep, then the viscosity of a material after a transformation will be smaller at lower temperatures.

sumed, proportional to the velocity of subduction. Under these assumptions, we solved the equation of the balance of the moment incorporating temperature, stress, and grain-size-dependent rheology of slab materials.

From such calculations, we can determine the rheological structure of subducting slabs and the resistance of such slabs against bending. The rheological structure of slabs from such an analysis is shown in figure 4-8. The results can be summarized as follows;

1. The rheology of a subducting slab is complex; even within the same slab, its central part has a very different strength and different deformation mechanisms compared to its surroundings.

2. Because nonlinear rheology becomes dominant with increasing

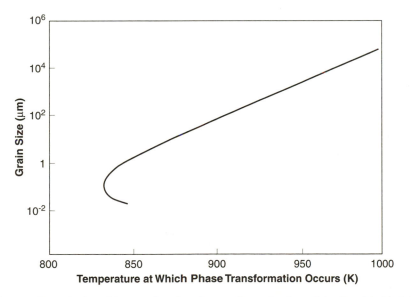

Fig. 4-7. The grain size of ringwoodite after the transformation in a subducting slab (Riedel and Karato 1997). Grain size is determined by the balance between the rates of nucleation and grain growth, and it is highly sensitive to the temperature at which the phase transformation occurs.

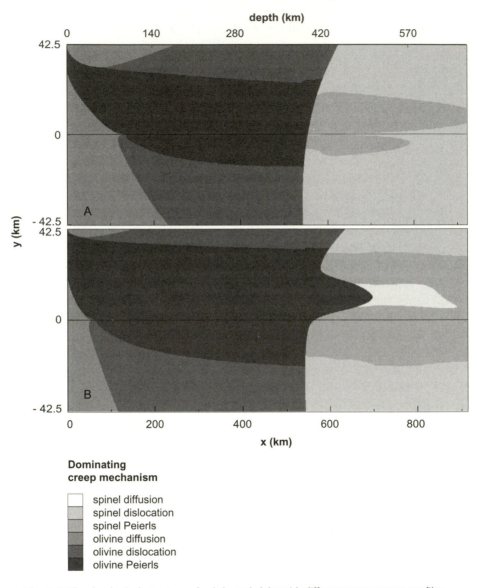

Fig. 4-8. The rheological structure of subducted slabs with different temperature profiles (after Karato et al. 2001). (a) Dominant mechanisms of deformation, (b) viscosity distribution.

Slab Viscosity Field

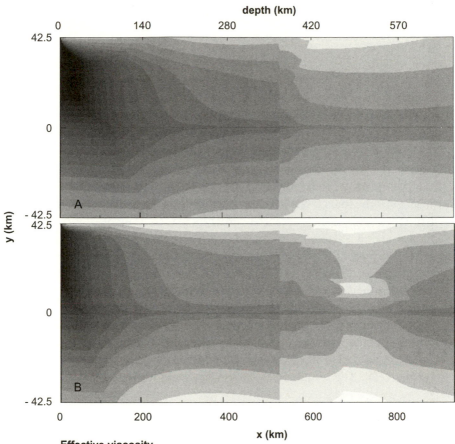

Effective viscosity

> 10^{40} Pas
$10^{38} \cdots 10^{40}$ Pas
$10^{36} \cdots 10^{38}$ Pas
$10^{34} \cdots 10^{36}$ Pas
$10^{32} \cdots 10^{34}$ Pas
$10^{30} \cdots 10^{32}$ Pas
$10^{28} \cdots 10^{30}$ Pas
$10^{26} \cdots 10^{28}$ Pas
$10^{24} \cdots 10^{26}$ Pas
$10^{22} \cdots 10^{24}$ Pas
$10^{20} \cdots 10^{22}$ Pas

stress, slabs cannot be so strong as expected from low temperatures (this was first noted by Goetze and Evans (1979)).

3. After a phase transformation, rheology becomes sensitive to grain size. Grain size is small, especially in the central part of cold slab after a phase transformation, and the viscosity in this region abruptly decreases.

4. Viscosity has a maximum just outside of this soft central part.

The resistance of such a slab against bending is measured by "flexural rigidity," defined by

$$D \equiv 4 \int_{-h/2}^{h/2} \eta(\dot{\varepsilon}) y^2 dy, \tag{4-2}$$

where h is the thickness of the slab, $\eta(\dot{\varepsilon})$ is the viscosity, and y is a distance from the center of slab measured perpendicular to its plane. It is more difficult to bend a slab with a larger D.

We have calculated D for slabs with various thermal structures (fig. 4-9). With a certain initial thermal structure of oceanic plate, its rheological structure is calculated for various subducting rates, and then the corresponding values of D are estimated. With the same initial temperature distribution, a greater subducting rate results in lower temperatures within a plate. Therefore, the results shown in figure 4-9 can be interpreted as a relationship between slab temperature (see box 4-2) and slab strength (D). The result can be summarized as follows.

1. The slab flexure rigidity defined by (4-2) is on the order of $\sim 10^{38}$ Nms. With $h \sim 100$ km, this would correspond to an average viscosity of $\sim 10^{23}$ Pa·s (compare this with an average viscosity of the deep mantle $\sim 10^{21}$–10^{22} Pa·s). So a slab is not as strong as one might expect from its low temperatures. This is caused by the effects of highly nonlinear rheology at high stress, cold regions (the Peierls mechanism; see box 4-3), and the effects of grain-size reduction.

2. There are two different trends in the relation between slab strength (D) and subducting rate (that is, slab temperature). When subduction is very slow and a slab is warm, D is small and the plate is soft. If the subducting rate increases from this state (so slab temperature decreases), D increases.

3. As the subduction rate increases and a slab gets colder, the effect of grain-size reduction due to phase transformation becomes more efficient. With this effect, lower temperature leads to smaller grain

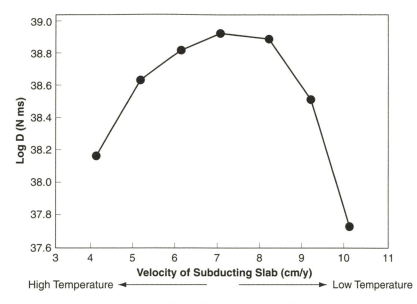

Fig. 4-9. The relation between slab flexural rigidity, *D*, at 600 km and at slab temperature (after Karato et al. 2001).

size, resulting in a smaller *D*. Also, a larger bending moment caused by a larger force at the 660-km boundary will cause the reduction of the strength of a slab with the subduction velocity. Consequently, when a slab temperature is sufficiently cold (i.e., has sufficiently large slab thermal parameters), a new trend appears in which *D* decreases with the subduction velocity (slab thermal parameter).

4. Because of these two trends, the variation of *D* as the function of the subducting rate (or slab temperature) has a maximum. Consequently, both warm and cold slabs are weak, but slabs with intermediate thermal structure are strong. Therefore, *the 660-km discontinuity may act as a rheological filter for mantle convection.*

The apparently paradoxical observation, that cold slabs are more deformed (softer) (fig.4-5), is naturally attributed to the temperature dependence of grain-size reduction after phase transformation(s) and to a higher stress caused by the faster subduction rate. When a slab is soft enough, then that slab can deform and could lie on top of the 660-km discontinuity. Trench migration observed in certain regions may be a result of slab weakening.

Some have proposed that as a rheological effect on the pattern of mantle convection, there is a viscosity reduction by "transitional superplasticity" in the transition zone. First-order phase transformations generate internal stress because of a volume change. This internal stress sometimes accelerates deformation, so it has been pointed out that viscosity might decrease in the transition zone. This is different from the effect of grain-size reduction discussed above, and it occurs only for nonlinear rheology—that is, dislocation creep where effective viscosity is stress-dependent. This phenomenon is well known in metallurgy, so the possibility was raised that a similar phenomenon might occur with earth minerals. Sammis and Dein (1974) suggested that viscosity at the depth of phase transformation might be globally low. Christensen and Yuen (1984) considered the effect of this low-viscosity zone and showed that layered convection was favored with this effect. A similar model was proposed by Panasyuk and Hager (1998). However, I consider that the notion of phase transformation superplasticity does not apply to most portions of Earth's mantle, and that viscosity reduction on a global scale is unlikely. This transformational superplasticity model can be tested by deformation experiments for a material under phase transition at high pressures. When my colleagues and I conducted deformation experiments on rocks under the olivine-spinel transition with a recently developed experimental technique, we could not detect any effect other than grain-size reduction. We also conducted similar experiments using analog materials at low pressures with a high-precession deformation device, and the results were the same. This is probably because internal stress is instantaneously released by dislocation recovery at high temperatures, so it does not enhance deformation. The same point was made by Paterson (1983).

4-2-3. The Separation of the Oceanic Crust from the Subducting Lithosphere

There is strong geochemical evidence that the materials of subducted oceanic crust are separated from the major portion of the mantle and that the basaltic magmas from ocean islands (e.g., Hawaii) originated from these regions (Hofmann 1997). Where could the subducted oceanic crust be separated from the main component of the lithosphere? Where is the source region for ocean island basalts (OIB)? Let us first consider the density structure of the subducting oceanic lithosphere, shown in figure 4-10. When the oceanic lithosphere is formed, its crustal part is composed of basalt, and its mantle part is composed of harzburgite, which consists of

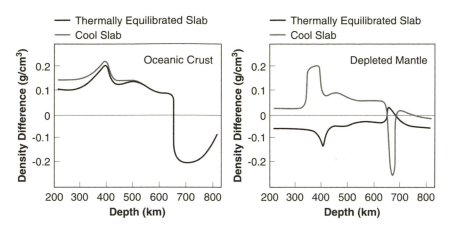

Fig. 4-10. The density difference between the subducted plate and the ambient mantle (pyrolite) (after Irifune and Ringwood 1987). Note that the crustal component becomes more buoyant than ambient mantle because it remains as garnet for 650–800 km.

mainly olivine and (ortho)pyroxene. Near the surface, the basaltic crust is much lighter than the mantle component. However, as the oceanic lithosphere subducts into the deep mantle, different phase transformations occur in these two portions. When the crustal part goes below ~ 400 km, it becomes rich in garnet and its density increases. The mantle part also becomes denser by other kinds of phase transformations, although not as dense as garnet. Around these depths, therefore, the crustal part becomes denser than the rest of the lithosphere, enhancing subduction. The situation becomes quite different in regions deeper than ~ 660 km. At this depth, the mantle part transforms into a very dense assembly composed of perovskite and magnesiowüstite. However, garnet will not transform to a denser structure (perovskite) until much higher pressure is attained. Thus, in the subducting lithosphere slightly deeper than 660 km—say, a 700-km depth—the crustal component must be more buoyant than the ambient mantle. At these depths, the main part of the subducting lithosphere has a higher density than the surrounding mantle (because of lower temperatures), but the subducted oceanic crust will be lighter than the surrounding mantle because of a large difference in densities. Irifune and Ringwood (1987) found that garnet-rich components retain a garnet-like structure until a depth of ~ 800 km. O'Neill and Jeanloz (1994) found that garnet-rich materials could survive to even deeper depth, ~ 1200 km.

Such a density relationship was predicted by Ted Ringwood (1982) and was experimentally verified (Irifune and Ringwood 1987). How does this kind of density structure affect subduction? Because the garnet component is very buoyant in the lower mantle, the density of the subducting lithosphere as a whole will be lower than that of ambient mantle at the depth of ~ 660–800 km (perhaps even deeper), even after including the effect of its lower temperature. Based on this result, Ringwood suggested that the subducting lithosphere could not easily penetrate the 660 km boundary, and that subducted materials might be accumulated above this boundary. He called this a *megalith*. This prediction was subsequently supported by the seismic tomography of Fukao, van der Hilst, and others, as discussed before (chapter 3).

Another possibility is that because the crustal component becomes buoyant at these depths, it may separate from the main part of subducting lithosphere and accumulate in the mantle transition zone (fig. 4-11). This idea was advocated by Ted Ringwood (1967, 1994) and Don Anderson (1989). If this actually happens, the chemical composition of the transition zone would be different from that of other parts, and it would also change with time. Though this is a very important issue for the chemical evolution of the Earth, it is difficult to test this model, and there is still no general consensus. Mark Richards and Geoff Davies (1989) and Jim Gaherty and Brad Hager (1994) made a numerical model of the deformation of the subducted oceanic lithosphere with a density structure similar to that shown in figure 4-10. Their models did not show crustal separation, so they concluded that the models of Ringwood and Anderson were unrealistic. They argued that this kind of small-scale density variation does not largely affect material circulation.

The subducted oceanic crust at this depth of ~ 600–700 km is made almost completely with garnet (Irifune and Ringwood 1987). Garnets are known to be much stronger than coexisting minerals. In a deformed garnet lherzolite (an upper-mantle rock containing olivine and orthopyroxene as well as garnet), although olivine (and orthopyroxene) show evidence of intense deformation, garnets are usually intact (they remain nearly undeformed). Therefore, it was obvious that the assumption that the subducted crust and the mantle components have similar viscosity was unrealistic. To further quantify the rheological contrast between garnet and other minerals, we have conducted an extensive study to quantitatively determine the rheological properties of a range of garnets (Karato et al. 1995b). The rheological structure predicted from these studies is shown in figure 4-12. Note in particular that the crustal component likely has a high viscosity, though it is not cold. If a large force is applied to this

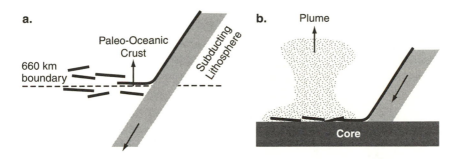

Fig. 4-11. The fate of the crustal component of the subducted oceanic lithosphere in the deep mantle. (a) In the depth range of ~ 660 – 1,000 km, the crustal component is still garnet, while ambient mantle is transformed into denser perovskite and magnesiowüstite. Thus, the crustal component is buoyant and may be separated from the denser portions of a slab. Karato (1997) showed that this separation is dynamically possible under certain conditions. (b) The crustal component may also reach the base of the mantle with subducting slabs itself and is separated from the main component (Christensen and Hofmann 1994). In this case, the crustal component is transformed into a dense phase composed mainly of perovskite, and it is denser than the ambient mantle because of iron enrichment.

Materials from the paleo-oceanic crust are important as source materials for ocean island basalts (OIB – i.e., hotspots). How the crustal component contributes to hotspots differs between the above two cases. In case (a), hotspots originate in the mantle transition zone (though a part of the heat may come from the lower mantle). In case (b), hotspots come from the core-mantle boundary.

thin and hard layer, it would appear that the layer will come off and be separated from the main component as was proposed by Ringwood and Anderson. A simple analysis was made to evaluate the conditions under which this delamination (separation) of oceanic crust might occur (Karato 1997). When a large force is applied, the garnet layer tends to separate from the main part of the subducted lithosphere through folding instability, if some materials easily flow in between the garnet layer and the main portions of lithosphere. Therefore, the separation of the garnet layer depends on the temperature and force as well as the viscosity of the garnet layer and the viscosity of surrounding materials. The most important conclusion drawn from this analysis is that the separation of the garnet layer, which was considered to be impossible based on models with uniform viscosity, can occur under reasonable mantle conditions if we consider the realistic rheological structure of the subducting lithosphere. We cannot understand the essence of small-scale dynamics, such as the sepa-

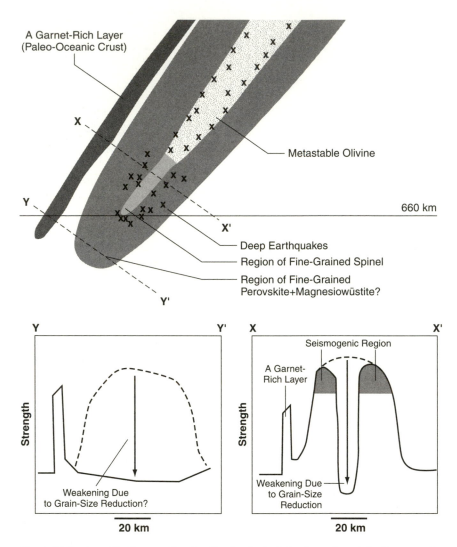

Fig. 4-12. A schematic diagram showing the rheological structure of a subducting oceanic lithosphere at around 600–1,000 km depth (Karato 1997). The effects of chemical layering are considered. The materials of the paleo-oceanic crust transform to garnets in this depth and therefore have great strength despite high temperatures. The main portion of a subducted lithosphere could become weak after the 660-km discontinuity because of significant grain-size reduction (Kubo et al. 1999).

ration of the garnet layer, without considering the fine-scale variation of viscosity.

Although these studies considered the separation of paleo-oceanic crust at the 660-km boundary, the separation depth does not have to be at 660 km. According to high-pressure experiments, the crustal component continues to be garnet, at least to the depth of ~ 800 km, or possibly to ~ 1,200 km (Irifune and Ringwood 1987; O'Neill and Jeanloz 1994); therefore, the separation of a paleo-oceanic crust might occur at greater depths. In fact, the recent experimental study by Kubo and his colleagues (1999) showed that the grain-size reduction associated with the ringwoodite → perovskite + magnesiowüstite transformation is very large, and hence the viscosity of the main component of subducted lithosphere in the (shallow) lower mantle will be small. In this case, separation of the paleo-oceanic crust might occur more easily in the shallow lower mantle. Seismic reflections seen at the depth of around 900–1,000 km may originate from separated oceanic crust.

What would be the consequence of the concentration of this garnet-rich material? If a large quantity of garnets are accumulated in a certain region, the viscosity of that region would increase so that it would become difficult to deform. Moreover, because these garnets are originally from the oceanic crust, this region has a significantly higher amount of radioactive elements compared to the ambient mantle. Therefore, the region of garnet-rich materials would be hotter but stronger than the ambient mantle: conditions suitable for the source regions of hotspots (OIB).

How can we then determine if this garnet-rich region exists? First of all, the density of garnet is very different from that of a standard model, so gravity may be affected. If garnet were present in the transition zone, its density would be higher than that given by a standard model. If garnet were abundant in the lower mantle instead, it would produce a negative density anomaly. We thus may be able to get some information from the observation of gravity and free oscillations. Furthermore, if a garnet layer were peeled off, garnet and the surrounding mantle might form a layered structure (e.g., Allègre and Turcotte 1986). Layering leads to elastic anisotropy, which causes anisotropy in seismic wave propagation. Therefore, we can also estimate the existence of garnet from seismic anisotropy. In fact, the study of free oscillations by Ishii and Tromp (chap. 3) found evidence for chemical heterogeneity around the transition zone, and it may indicate the existence of a large amount of garnet. In addition, the anisotropy of the lower transition zone and the shallow lower mantle (Wookey, Kendall, and Barruol 2002) cannot be easily explained by the preferred orientation of minerals, and it may result from the layering of garnet and the surrounding mantle.

Additional evidence is of geochemical (petrological) nature. Based on the analyses of the chemical compositions of basalts, there is a growing consensus that materials rich in paleo-oceanic crust are involved in the generation of basalts, including MORB, but more prominently in basalts from hotspots (OIB). Eiichi Takahashi, at the Tokyo Institute of Technology, and his colleagues proposed that some of the hotspot magmas involve the melting of a mixture of typical mantle materials with paleo-oceanic crust coming, presumably, from the transition zone (Takahashi, Nakajima, and Wright 1998).

In summary, I conclude that if realistic rheological properties are considered, the separation of the paleo-oceanic crust is possible under some conditions at around ~ 660–1,000 km depth as well as at the base of the mantle, as proposed by Christensen and Hofmann (1994). Although the dynamics are quite different, the separation of the lower continental crust through gravitational instability has also been proposed (Meissner and Mooney 1998). Once separated, these components will be trapped at around the 660-km discontinuity. Therefore the source regions of OIB can be either the region around the 660-km discontinuity or the region at the bottom of the mantle (the D″ layer), or both.

4-3. THE INFLUENCE OF RHEOLOGY IN MIXING

There are several possible processes by which different chemical reservoirs may have been formed. Chemically distinct materials such as paleo-oceanic crust ("undepleted") or the continental crust ("enriched") and the MORB source mantle ("depleted") component may be separated from each other in the manner described above. Alternatively, crystallization in a putative magma ocean could cause chemical layering (e.g., Ohtani 1985; Agee 1993). Once formed, how could these regions survive in a dynamically convecting mantle without being mixed up with the surrounding materials? This question becomes particularly important because there is strong evidence that the whole mantle is involved in convection, as discussed in section 4-1. The simplest model that Earth's mantle is layered and that there is no material exchange between these layers is no longer tenable. Acceptable models for chemical evolution must be consistent with (present-day) nearly whole mantle convection and the presence of long-lived chemical reservoirs (for at least one billion years). As we will see, this is a challenging problem, and there is no widely accepted model that is consistent with geophysical and geochemical observations as well as with the properties of materials composing Earth.

The processes and efficiency of mixing have been examined from the geodynamic point of view by several geodynamicists. Louise Kellogg and Don Turcotte, then at Cornell University, for example, investigated the mechanisms of convective mixing through stretching and chemical diffusion (1986–87). They considered that two materials will be chemically mixed when the characteristic length of each component becomes smaller than the length scale of atomic diffusion for a given time scale. Based on the values of diffusion coefficients of various atomic species determined in the laboratory, the characteristic length for diffusion is estimated to be on the order of ~ 10 cm for billions of years. Their analysis showed that the time to deform these regions to a thickness smaller than this critical value is inversely proportional to the average strain rate, and for the strain rate of 10^{-15}–10^{-16} s^{-1}, the time frame is ~ 0.2–2 billion years: the efficiency of chemical homogenization by convection-induced deformation is marginal. Layered structures with a range of thicknesses (~ 1 cm to ~ 1 m) are often found; in this case, homogenization would have occurred for only thin layers.

Three points should be noted here.

1. The degree of homogenization depends on the processes of chemical diffusion. Kellogg and Turcotte considered chemical diffusion through the solid state. Mass transport through melt occurs much faster, particularly when the advection of melt occurs. In most cases, melt in the upper mantle assumes continuous tubules along grain boundaries (chap. 2); hence, long-range melt transport through these tubules is a plausible mechanism of melt transport. In these cases, a much broader region could attain chemical equilibrium.

2. The above analysis is for steady-state, two-dimensional flow. In a real Earth, where convection is vigorous and chaotic, three-dimensional flow will mix materials more effectively (Christensen and Hofmann 1994; van Keken and Zhong 1999). These studies showed that homogenization is more effective in these cases, and that the time-frame for homogenization is on the order of several hundred million years.

3. Although this point has been ignored quite often, the efficiency of mixing is highly dependent on the rheological contrast of two materials. Michael Manga, then at Harvard University, was the first to investigate the importance of rheological contrasts on mixing (Manga 1996). Manga considered the deformation of blobs embedded in a matrix (fig. 4-13a).

The densities of the two materials are assumed to be the same. Using an analytical solution of two-dimensional flow, Manga showed that if blobs embedded in a matrix have significantly higher viscosity than the matrix, then the convective mixing is much less efficient, and those blobs tend to aggregate to make a large "reservoir" (1996). Norman Sleep, at Stanford University, investigated the efficiency of the entrainment of dense materials from the convection current above them (1988) (fig. 4-13b). Sleep showed that the degree of entrainment depends on the density and viscosity contrasts of two materials. If the viscosity is the same, then the density of the lower layer must be higher than the mantle above by ~ 6% or more to be isolated from the main circulation for several billions of years. However, if the viscosity of the dense layer is higher than the layer above, then the efficiency of entrainment decreases. Both studies showed that if the materials in a layer or the blobs have significantly higher viscosity than surrounding materials, then they could survive more effectively than in the case of a homogeneous viscosity. If the viscosity of materials is higher than the surrounding materials by a factor of 10–100, then these high viscosity materials can be isolated from the main circulation for several billion years. The situation is quite similar to the case of the longevity of the continental tectosphere: the continental tectosphere has survived for billions of years presumably due to its high viscosity caused by the removal of water (Pollack 1986) (chap. 2).

Several conclusions can be drawn from these studies. First, the processes of mixing are complicated because they depend on the scale of homogeneity, the geometry, and the physical properties of the materials to be mixed. Consequently, it is likely that the degree of heterogeneity in Earth's mantle has a diverse spectrum depending upon the scale of heterogeneity (e.g., Hart 1988). Second, however, these analyses show the difficulties inherent in maintaining large-scale heterogeneities even in a layered structure, as far as whole-mantle convection occurs. In particular, it is notoriously difficult to isolate "undepleted" or "enriched" materials from the general circulation in the mantle, because these materials, by definition, contain a larger amount of radiogenic elements and will be warmer. In addition, these materials are likely to have more volatile components, such as water. These two factors likely reduce the viscosity of these materials and, as a consequence, they would easily be mixed with the surrounding materials by convection.

Recognizing these difficulties, Thorsten Becker and his colleagues at Harvard discussed possible mechanisms for the survival of geochemically

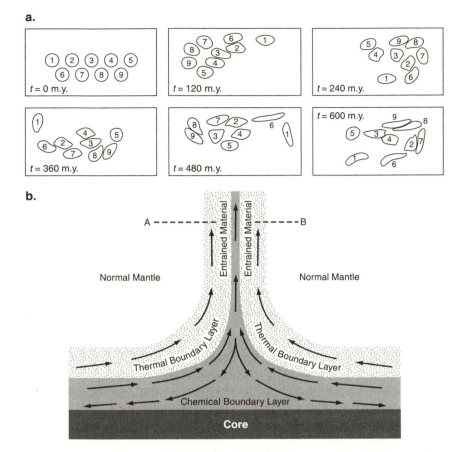

Fig. 4-13. (a) Mixing due to deformation of blobs by convection. Stretching by deformation causes thinning of a blob until its thickness reaches some threshold, beneath which chemical diffusion will homogenize the composition. The rate of deformation of blobs depends strongly on their viscosity, and when they have significantly higher viscosity than the surrounding materials, they are effectively isolated from the major convection circulation (Manga 1996). (b) Mixing by entrainment (covectional erosion). Materials in a dense layer at the bottom of the mantle can be entrained by the convection current in the above layer. The rate of entrainment depends on the density contrast as well as viscosity contrast (Sleep, 1988).

"enriched" (or "undepleted") reservoirs (Becker, Kellogg, and O'Connell 1999). Following the idea by Michael Manga, they tried to find conditions under which geochemically "enriched" or ("undepleted") reservoirs are equally as dense as the surrounding materials and yet have a higher viscosity. They considered that blobs with "enriched" (or "undepleted") compositions have a higher concentration of radiogenic elements than do the "depleted" surrounding regions, and asked if a condition is met under which no significant mixing occurs. Briefly, this analysis underscored the difficulties "enriched" (or "undepleted") blobs to survive. The most fundamental problem is that these blobs are considered to have a higher concentration of heat sources, thus tending to warm themselves, and it is hard to make them stronger than the depleted regions. These authors proposed that ancient "enriched" (or "undepleted") blobs would have a higher concentration of perovskite, which might lead to a high viscosity. Such an idea is consistent with the current knowledge of the rheology of lower-mantle materials (Yamazaki and Karato 2001b), but a large difference in chemical composition likely causes a detectable difference in acoustic impedance. There is no clear evidence for a sharp seismic discontinuity in the lower mantle, except at the top of the D″ layer and in small-scale scatters (Hedlin, Shearer, and Earle 1997). Similarly, the model by Kellogg, Hager, and van der Hilst (1999) has difficulty explaining the mechanisms of the survival of the layered structure because the hypothetical deep undepleted reservoir is enriched with radiogenic elements and hence likely to have a low viscosity.

How could we explain the long-term survival of "enriched" or "undepleted" geochemical reservoirs? A possibility is to assume that the convection pattern has changed through geological time, and that the style of mantle convection was two-layered in the past and that the seismic evidence of whole-mantle convection applies only to the recent Earth. Claude Allègre proposed this idea to explain the isotopic anomalies of Ar (1997). The change from layered to whole-mantle convection is plausible on a geodynamic basis (e.g., Christensen and Yuen 1985; S. Honda 1995); this is still a viable model, although its test is not straightforward. Similarly, Paul Silver and others suggested a hybrid version of a mantle convection model to reconcile geochemical observations with geophysical observations (Silver, Carlson, and Olson 1988).

There is another theoretical, albeit highly speculative, possibility. "Enriched" or "undepleted" regions could indeed have higher viscosity than "depleted" regions. The majority of the lower mantle is likely to deform by diffusion creep (Karato, Zhang, and Wenk 1995). When materials deform by diffusion creep, the viscosity is highly sensitive to grain size. The grain size of materials deforming by diffusion creep is controlled by grain-growth kinetics, which depend on the temperature (and water content). Consequently, if the temperature dependence of grain size is high, then the higher is the temperature, the larger the grain size. As a result, the viscosity of materials could become *higher* at high temperatures. A similar case was discussed in section 4-2, where grain size is controlled by temperature through nucleation and growth processes associated with phase transformations. Such an anomalous temperature dependence of viscosity was also discussed by Slava Solomatov (1996).

Finally, if the "enriched" or "undepleted" regions occur in the transition zone or the upper regions of the lower mantle, these materials are composed mostly of garnet. In this case, they would naturally have a higher viscosity than the surrounding materials and, hence, would survive against convective erosion.

4-4. SUMMARY AND OUTLOOK

The pattern of mantle convection and the processes of chemical differentiation in Earth are obviously dependent upon properties of Earth materials. Among others, densities and viscosities are most important properties controlling these processes. However, comparing to density, measurements of the viscosity of Earth materials under deep Earth, conditions are more difficult. Consequently, we have very limited knowledge of the rheological properties (viscosities) of deep Earth materials. Presumably for this reason, most of the previous studies on these geodynamic processes ignored the role of rheological properties or treated it in only a simplified way. As I have tried to demonstrate in this chapter, such a too-simplified approach could miss some important points. Obviously, however, much of the discussions on the influence of rheological properties presented in this chapter remain highly speculative, mostly because of the many uncertainties in rheological properties. Such a situation is rapidly changing because of new developments in experimental techniques of high-pressure deformation. In the next generation of geodynamic studies, some key aspects of the rheological control of geodynamic processes must be included to improve our understanding of the dynamics and evolution of this planet.

FIVE • THE ORIGIN
OF DEEP EARTHQUAKES

5-1. THE DISCOVERY OF DEEP EARTHQUAKES

Earthquakes usually occur at shallow depths in Earth. The relatively shallow part of Earth stays at low temperatures, and because the strength of rocks there is so great, a large amount of elastic strain energy can accumulate. If certain conditions are met, rocks fail by brittle fracture, releasing a large amount of elastic energy as waves. This phenomenon is called an *earthquake*. At greater depths in Earth, however, temperature and pressure are both high, so rocks must deform by ductile flow, not by brittle fracture, when external force is applied. Kiyoo Wadati, at the Central Meteorological Observatory in Japan, investigated the locations of earthquakes using the differences in P- and S-wave arrival times and realized, contrary to expectations, that some of the earthquakes must be located deeper than ~ 300 km (Wadati 1927). His report was received with great skepticism at first. Harold Jeffreys, in the United Kingdom, an authority on geophysics at that time, did not readily believe the existence of deep earthquakes, and argued that the high stresses required to cause earthquakes cannot exist in the deep Earth because materials can flow easily there, according to the principle of isostasy. Wadati persistently continued his research and conducted an extensive follow-up within a few years of his initial discovery, finally succeeding in obtaining worldwide recognition for the presence of deep earthquakes. His follow-up works include the demonstration of the planar regions of deep earthquake activities (now known as the Wadati-Benioff zone), and the correlation of this zone with gravity anomalies and with regions of anomalous seismic vibrations (i.e., regions of low attenuation). It was more than twenty years later when Hugo Benioff, at Caltech, pointed out the presence of a similar deep-earthquake zone and discussed its origin (Benioff 1949). A deep-earthquake zone is now called as the Wadati-Benioff zone, but it had been often called simply the Benioff zone until the mid 1970s. Wadati's achievements have finally attained universal recognition, thanks to the efforts of a number of Japanese scientists.

Once the existence of deep earthquakes was widely accepted, their significance for tectonics was quickly recognized. Arthur Holmes, in Britain, for example, immediately related the deep seismic zone (the Wadati-Benioff zone) with mantle convection. During the development of the theory of plate tectonics, the Wadati-Benioff zone played a key role in defining the concept of the *subduction* of a plate into the deep mantle. From the beginning of their discovery, however, a number of enigmatic issues remain regarding the physical mechanisms of deep earthquakes. Even more than seventy years after the discovery, we do not yet have a widely accepted theory for the origin of deep earthquakes. In this chapter, starting from the basics of earthquakes, I will explain our current understanding of the mechanism of deep earthquakes.

5-2. THE CHARACTERISTICS OF DEEP EARTHQUAKES

One of the important findings since the discovery by Wadati is that deep earthquakes occur within a subducting slab, which is under high pressure but at a relatively low temperature. We now know that, contrary to Jeffereys's argument, deep earthquakes do not occur in regions where temperatures are high.

Furthermore, it has also been found that many deep earthquakes occur where phase transformations of minerals are likely to occur. High-pressure mineral physics studies (see chap. 1) contributed to this discovery. Not all of deep earthquakes occur, however, at depths where phase transformation may occur. Events at ~ 70–300 km are sometimes called intermediate-depth earthquakes, and some people classify them differently from deeper events at ~ 300–680 km (which are deep earthquakes in the strict sense). As will be explained shortly, intermediate-depth earthquakes are equally as enigmatic as deep earthquakes. It is therefore desirable if a model for deep earthquakes can also explain intermediate-depth earthquakes in a unified manner. Major constituent minerals do not undergo phase transformations at these depths, and, as will be shown below, there is no clear relation between the depth distribution of earthquakes and phase transformations. The relation between deep earthquakes and phase transformation is thus not expected to be so straightforward.

Major steps taken in the study of the mechanism of deep earthquakes include the recognition of the tectonic significance of the Wadati-Benioff zone, advances in mineral physics based on high-pressure experiments, and improved seismic observations. In the late 1980s, in particular, experimental studies of instabilities associated with phase transformation were conducted by Steve Kirby's (at the U.S. Geological Survey) and Harry Green's (then at the University of California at Davis) groups, and the kinetics of phase transformation were investigated by Dave Rubie and others. A few years after these studies, the greatest deep earthquake on record occurred (the Bolivian Earthquake, with a magnitude of 8.3), and the detailed observation of this earthquake contributed to our improved understanding of deep earthquakes. A satisfactory model for deep earthquakes, however, has not emerged. The models so far proposed do not explain seismic observations very well, or they lack a sound physical basis. In this chapter, I will critically review the existing models and explain current research efforts toward a new model.

5-2-1. Distribution

The depth distribution of earthquakes is shown in figure 5-1. From the surface to a depth of ~ 300 km, the frequency of earthquakes monotoni-

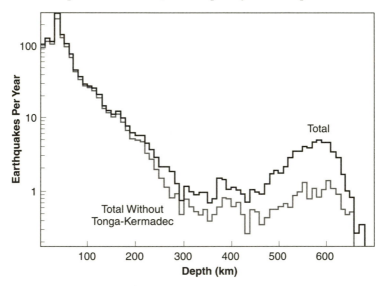

Fig. 5-1. Depth distribution of earthquakes. From the surface to ~ 300 km depth, the frequency of earthquakes rapidly (exponentially) decreases with depth. In regions deeper than ~ 500 km, however, seismic activity increases again and peaks at ~ 600 km. No earthquakes have been detected deeper than 680 km. Most of intense seismic activities at ~ 600 km depth take place in the Tonga slab (Green and Burnley 1989).

cally decreases with depth. It peaks at ~ 550–600 km, and earthquake activity completely stops at ~ 680 km depth. Events at deeper than 200–300 km are seen only where the oceanic plate is believed to subduct into the deep mantle (very rarely, some deep earthquakes are observed far from the inferred region of the subducted lithosphere. These are often referred to *isolated earthquakes,* which may represent the location of fragmented paleo-subducted slabs). The activity of deep earthquakes is highly spatially variable. About two-thirds of the events at deeper than ~ 500 km, in particular, take place at the Tonga-Kermadec subduction zone in the south Pacific. There is a similar relation between the depths and magnitudes of earthquakes. Though the depth distribution is not quite clear, there is a tendency for large deep earthquakes to occur at great depths. The focal depth of the Bolivian Earthquake of 1994 (magnitude 8.3), for example, was 637 km. On the other hand, the Tonga-Kermadec subduction zone, where a number of deep earthquakes take place, does not have many giant earthquakes.

5-2-2. Focal Mechanisms

By studying the pattern of seismic wave generation, Hirokichi Honda in Japan showed that deep earthquakes were caused by faulting (a sudden shear motion along a plane), as is the case with shallow earthquakes (1932). Faulting is a localized deformation that occurs as a result of the interaction of number of cracks. Under high confining pressure and at high temperatures, rocks usually deform by ductile flow. When rocks deform by ductile flow, localized deformation such as faulting does not usually take place. In deep Earth, ductile deformation is expected to be dominant because of the high temperature and great pressure, so that the faulting origin of deep earthquakes was once questioned. Like an underground nuclear explosion, an abrupt volume change associated with phase transformations was considered to be the source mechanism of deep earthquakes (Bridgman 1945). Through the detailed analysis of the pattern of seismic waves generated from deep earthquakes, however, the possibility of a sudden volume change has been ruled out. Deep earthquakes, just like shallow earthquakes, occur by shearing motions on a fault.

The stress drop associated with an earthquake can be calculated from the size of a source region and the amount of energy released by an earthquake (the latter is measured by "seismic moment"). The stress drop is the difference between the stress acting on a fault plane before and after an earthquake, so the actual stress must be larger than the observed stress drop. The stress drop associated with deep earthquakes tends to be larger than that associated with shallow earthquakes. From this observation, the stress within a subducting plate is found to be larger than the stress within a plate near the surface.

Another important factor regarding focal mechanisms is the direction of first motions—that is, the direction the motion of materials in a seismic source. This has been traditionally studied for many years, and it is known that for events deeper than 300 km, the axis of compression tends to be parallel to the direction of subduction. The pattern of first motions is, however, more complicated for intermediate-depth earthquakes. In the well-studied case of northeastern Japan, the first-motion pattern is systematically different between earthquakes at the top surface of the plate and those within the plate; the top surface seems to be in compression, while the inside seems to be in extension.

5-2-3. Aftershocks

One of the major characteristics of deep earthquakes is that the number of aftershocks is extremely small compared to those for shallow earthquakes. In the case of the Colombian Earthquake of 1970 (magnitude 7.6, focal depth of 653 km), for example, no aftershock was detected at all. Doug Wiens, at Washington University in St. Louis, and others showed that the aftershock zone usually extends over almost an entire subducted slab, and that aftershock activities are prominent only in very cold slabs (Wiens and Gilbert 1996). They also noticed that the relationship between the number of aftershocks and their frequency depends on the temperatures of the subducting slab where an earthquake occurs. The number of aftershocks, N, is related to the magnitude of aftershocks, M, as $logN = a - bM$, where the values of a and b are parameters. According to Wiens and others, the b value of aftershocks is smaller for deep earthquakes in a warmer plate (Wiens and Gilbert 1996) (fig. 5-2). A similar report was also made by Cliff Frohlich (1989).

5-2-4. Efficiency

The efficiency of an earthquake is defined as the ratio of the energy released as seismic waves to the elastic strain energy accumulated before an earthquake. By analyzing the Bolivian Earthquake, Hiroo Kanamori, at Caltech, and others pointed out that it had a lower efficiency and a slower rupture propagation than shallow earthquakes do (Kanamori, Anderson,

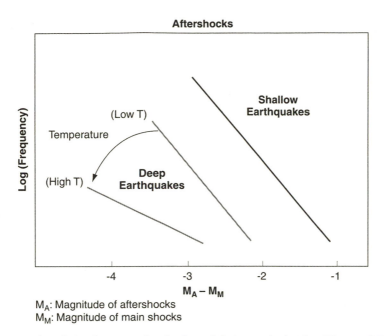

Fig. 5-2. The relation between aftershocks and their magnitude (after Wiens and Gilbert 1996). Deep earthquakes generally have fewer aftershocks than shallow earthquakes. This trend is more clearly seen for warmer slabs. Small aftershocks are especially rare for deep earthquakes in warm plates.

and Heaton 1998). According to their calculation, the efficiency of the Bolivian Earthquake was 3.6%, and its rupture velocity was only 20–30% of shear wave velocity (the efficiency of shallow earthquakes is usually greater than 10%, and the rupture velocity is 70–80% of shear wave velocity). This indicates that the rupture was rather viscoelastic, unlike a perfect brittle fracture. Wiens and others examined this point in more detail and showed that this trend is most prominent for warm slabs (i.e., young and slowly subducting).

5-3. ROCK MECHANICS AND EARTHQUAKES: BRITTLE-DUCTILE TRANSITION

To understand why deep (and intermediate-depth) earthquakes are so enigmatic, I will briefly explain how normal, shallow earthquakes are considered to be generated. An earthquake is a phenomenon in which (elastic) energy accumulated within Earth is released suddenly as elastic waves. For an earthquake to occur, therefore, large elastic energy has to be stored in the first place. Since elastic energy per unit volume can be expressed as (stress)^2/(elastic modulus), this condition is equivalent to the existence of sufficiently large stress. At shallow depths, this condition is easily satisfied because rocks have enough strength to support great stress. In the hot mantle in the deep portion of Earth, rocks cannot support such a large stress. With the viscosity of 10^{21} Pa·s the stress that can be supported for a typical geological time scale (say, ~ 10 million years) is $\sim 10^6$ Pa, corresponding to the elastic strain energy (per unit volume) of only ~ 10 J/m^3, which is too small to generate an earthquake (the energy density of $\sim 10^6$ J/m^3 is required to release an earthquake with the moment of $\sim 10^{21}$ N m (magnitude ~ 7.5) from the volume of $\sim 10^6$ km^3). This difficulty can be easily resolved by considering that a subducting slab has a viscosity higher than the ambient mantle. With the viscosity of 10^{24} Pa·s, the stress that can be supported for ~ 10 million years is as large as 10^9 Pa, corresponding to the energy density of $\sim 10^7$ J/m^3. Therefore, by noting that deep earthquakes occur in anomalously cold regions of Earth, the first problem of earthquake generation is easily solved.

A more difficult problem is to understand a physical mechanism by which this strain energy is released by abrupt fault motion at high confining pressures. Abrupt fault motions are possible under the following conditions. Consider a case of faulting due to the interaction of cracks. Cracks play an important role in faulting and fracturing under low pressures. Any rock is heterogeneous, inherently with innumerable microcracks. When an external force (stress) is applied, these cracks propagate and grow by interacting with each other. Crack propagation is driven by stress concentrated on the tip of a crack, and stress concentration becomes greater for a larger crack. Thus, the deformation of rocks due to crack propagation is essentially unstable. If some part happens to deform more than other parts, it continues to deform more and more easily, leading to runaway deformation—that is, fracture. Fracture, therefore, tends to be localized in a certain plane.

Under high pressures, however, this type of fracture occurs with difficulty. Because crack formation (and propagation) is associated with an increase in the volume (called dilatancy), cracks are difficult to form and propagate at high pressures. The final stage of a fracture—that is, a slip along a fault plane—is also difficult to achieve at high pressures. To understand this more quantitatively, let us consider the resistance for the sliding along a fault plane. This resistance is basically *friction* for which the following law (Coulomb-Navier's law) is known:

$$\sigma = \sigma_o + \mu\sigma_n, \tag{5-1}$$

where σ is the shear stress needed to move the fault, σ_o is the cohesive strength (fault strength at zero normal stress), σ_n is the normal stress, and μ is the friction coefficient. By compiling a large number of experimental results, Jim Byerlee, at the U.S. Geological Survey, noticed that the parameters σ_o and μ in the above equation were almost constant, regardless of the type of rocks. This is called the Byerlee's law. From this relation, it can be readily seen that slip along a fault is difficult under high pressures, because normal stress increases in proportion to pressure. For the pressure of 24 GPa, which corresponds to the depth of 650 km, a differential stress of about 20 GPa is required to cause faulting. No tectonic process can produce such a high differential stress, so we can conclude that faulting due to brittle fracture is impossible at this great depth. Even for the depth of ~ 50 km, the required differential stress for a brittle fracture is as high as 1 GPa.

Ductile deformation is another mechanism of deformation (boxes 1-5, 1-6). The strength of the material corresponding to ductile deformation may be written as (from equation [B1-6-2]),

$$\sigma \propto exp\left[\frac{H_\eta^*}{nRT}\right]. \tag{5-2}$$

This strength decreases rapidly with increasing temperature, so the ductile strength is small in the relatively deep mantle. At a certain depth within the Earth, therefore, the deformation mechanism changes from brittle fracture to ductile deformation. This is called the brittle-ductile transition (box 5-1). Though this argument must be slightly modified if we consider the effect of pressure on ductile deformation, it is still true that ductile deformation takes place more easily than brittle fracture in the deep mantle. The question is, then, why faulting is ever possible in the deep mantle, where ductile deformation should dominate.

5-4. THE MECHANISMS OF DEEP EARTHQUAKES

5-4-1. The Origin of Stress

Great stress is required to generate an earthquake. For a subducted slab, there are a number of sources for such a large stress. The subducted slab itself is pulled down by gravity because it is denser than the surrounding mantle. The focal mechanisms of deep earthquakes in subducting slabs

Box 5-1. Earthquakes and the Brittle-Ductile Transition

The deformation of rocks can be classified into *brittle* and *ductile* deformations. Brittle deformation is deformation by fracture that occurs at low temperatures and low pressures. Microscopically, brittle deformation occurs through the propagation of cracks. Cracks interact each other and often make a *fault*. Therefore, the resistance of a material to brittle failure is controlled by the resistance to fault motion. The resistance for fault motion is proportional to the confining pressure, and fault motion (brittle failure) becomes difficult at great depths. Ductile deformation refers to deformation by plastic flow. Microscopically it occurs through the motion of *lattice defects* (box 2-2). High pressures suppress brittle failure and high temperatures promote the motion of lattice defects. Consequently, in the deep interior of Earth, where both pressure and temperature are high, ductile deformation is the dominant mode of deformation, and earthquakes do not usually occur (fig. B5-1-1). In most cases, ductile deformation occurs uniformly, but when certain conditions are met, localized deformation can occur, leading to unstable, faultlike deformation.

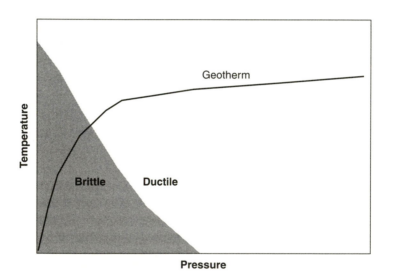

Fig. B5-1-1. A schematic diagram showing a transition from brittle to ductile deformation in a pressure-temperature space. Both pressure and temperature are high in Earth's deep interior and ductile deformation is expected.

show that subducting slabs encounter some obstacles at great depths. A density jump and/or a viscosity increase at the 660 km discontinuity are the possible causes for such resistance. Density variation due to phase transformations as well as thermal stress can also be sources for stress. In addition to these forces that may act in the deep portions of subducting slabs, elastic bending (near trenches) also causes large stresses that vary within the thickness of a plate. In any case, these stress sources must be consistent with observations in terms of stress orientation and magnitude. Except for the stress due to plate bending, almost all of the above mechanisms predict compressional stress parallel to the subduction direction, so they are consistent with seismic observations. For intermediate-depth earthquakes (~ 70–300 km), double seismogenic zones are sometimes observed, and these two zones have different focal mechanisms. This can be explained by the effects of plate bending.

5-4-2. The Mechanism of Fault Formation

Why do these stresses cause abrupt faulting, not ductile flow? This is a much more difficult question to answer than identifying sources of stresses, and we do not have a generally accepted hypothesis yet.

Barry Raleigh and Mervyn Paterson showed that the breakdown of serpentine, one of the hydrous minerals, could lead to the formation of a fault (1965). They interpreted the data to explain that fault formation resulted from the pressure of water released by the breakdown of serpentine. When fluid is present on a fault plane, the effective normal stress in equation (5-1) becomes σ_n-P_f (P_f stands for fluid pressure). An increase in fluid pressure lowers this effective pressure, and the fracture strength of rocks decreases as well. They suggested that this process may also occur in deep mantle and produce deep earthquakes. Because the breakdown of serpentine and other hydrous minerals does not occur deeper than ~ 200–300 km, it is difficult to explain deeper earthquakes with this hypothesis. In addition, whereas a large amount of hydrous minerals may be present on the surface of subducting slabs, they are not expected very much within a plate. According to the study of hydrothermal circulation, seawater is considered to penetrate only to ~ 4–6 km into the oceanic plate. Therefore, even for the relatively shallow events in deep earthquakes (shallower than ~ 200–300 km), this mechanism can explain only those that occur at the top surface of a plate. It is difficult to account for those that occur deeper in a plate.

Recently, a few groups of scientists have emphasized the importance of a dehydration reaction as a cause for intermediate earthquakes, including those which occur inside of subducting slabs (intermediate earthquakes in the deeper zone). Seno and Yamanaka noticed that the location of a double seismic zone of intermediate earthquakes coincides with the conditions for the dehydration of serpentine, and proposed that subducting slabs are hydrated to at least ~ 30–40 km from their surface and that the dehydration of serpentine is responsible for intermediate-depth earthquakes (Seno and Yamanaka 1996). A similar model was proposed by Simon Peacock, at Arizona State University (2001). A main problem with these models is the difficulty in explaining the hydration of subducting slabs. Seno and Yamanaka (1996) suggested that the hydration by magmas from hotspots as a mechanism, and Peacock (2001) suggested that water might penetrate along the normal faults at the outer region of oceanic trenches. Hydration by hotspot volcanism is unlikely to explain the wide distribution of intermediate-depth earthquakes (it would cause only a localized hydration). Similarly, it is not clear how water can penetrate deep into the slab during the faulting that usually initiates in the deeper portion and propagates upward.

Phase transformations in a subducting plate have also been suggested as a cause for deep earthquakes. One attractive point of this idea is that the depth distribution of deep earthquakes has some correlation with the depths of phase transformations. In particular, the phase transformation model is consistent with the absence of seismic activities deeper than 680 km, where phase transformation also ceases to occur. However, when we examine this idea in more detail, we see that the correlation between deep seismicity and phase transformation is more complicated. If all phase transformations take place at equilibrium conditions, the depths of phase transformations cannot explain the depth distribution of deep earthquakes, which is shown in figure 5-1. Seismic activity is the lowest at a depth of ~ 350–520 km, where the olivine-wadsleyite or olivine-ringwoodite transitions would take place in equilibrium. Several scientists have proposed that the relation between phase transformation and earthquakes may be attributed to nonequilibrium, delayed-phase transformations (see chap. 1). The low temperatures of a subducting plate (the center of the plate is colder than the ambient mantle by 500–1,000 K) could cause delayed phase transformation. Thus, the olivine-wadsleyite and olivine-ringwoodite transitions could take place at around 500–650 km in a subducting plate, more than ~ 200 km deeper than the depths for equilibrium transformation.

Percy Bridgman at Harvard, the pioneer of high-pressure physics (he received the Nobel Prize for physics in 1947), advocated that phase transformations might be the origin of deep earthquakes (Bridgman 1945). In this early model, however, he suggested that an abrupt volume change due to phase transformations could generate seismic waves. This hypothesis, however, is not considered to be viable because the analyses of the focal mechanisms of deep earthquakes clearly demonstrate that there is no appreciable volumetric strain associated with deep earthquakes. According to this idea, the focal mechanism of deep earthquakes, unlike that of shallower earthquakes, must have an isotropic compressional component, which is not consistent with observations.

Consequently, most people thought for a long time that phase transformations could not cause deep earthquakes. In 1987, Steve Kirby, at the U.S. Geological Survey, however, revived the phase transformation model with a new twist (1987). By conducting experiments on materials that undergo phase transformations, such as ice, he found out that if deviatoric stress is applied during a phase transformation, some materials deform unstably to generate elastic waves. According to his experimental results, this instability is not associated with a sudden volume change, but is caused by an unstable shear deformation that leads to the formation of a fault. Unlike Bridgman's model, this does not contradict seismological observations. Kirby's analysis shows that this type of instability is limited to exothermic phase transformations with a large volume change. This condition is satisfied by both the olivine-wadsleyite and olivine-ringwoodite transitions, so he considered that this mechanism might be responsible for deep earthquakes (Kirby 1987; Kirby, Durham, and Stern 1991). Shortly after this proposal, Harry Green also presented a similar model. By conducting low-pressure (1–2 GPa) experiments using an analog material, Mg_2GeO_4, Green more directly demonstrated faulting instability caused by the olivine-spinel transition (Green and Burnley 1989). Green also conducted ultra-high-pressure (~ 15 GPa) deformation experiments and reported fault formation due to the silicate olivine-wadsleyite transition, suggesting this phenomenon as a likely mechanism for deep earthquakes (Green et al. 1990; Green 1994). This type of model is called the *transformation-faulting model*.

How can phase transition lead to faulting? According to Kirby and Green, one possible physical model may be described as follows (fig. 5-3). Consider a phase transformation from phase A (olivine) to phase B (spinel). Phase B is assumed to be denser than phase A. With increasing pressure, phase A is transformed into phase B. This transformation does not occur abruptly; phase B is first nucleated as a small lens-shaped nu-

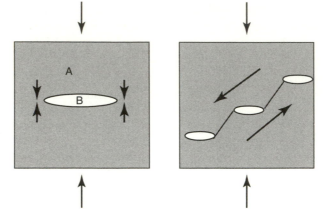

Fig. 5-3. Mechanism of transformation faulting (after Green 1994). Because of a volume change upon the transition from low-pressure phase A to high-pressure phase B, stress focuses on the tips of phase B. This compressional stress enhances the transition to phase B, so this phase grows. This process is similar to crack propagation. Eventually, a number of phase B lenses are connected to result in faulting.

clei, and phase transformation is achieved by its gradual growth. When differential stress is applied, the lens-shaped phase B tends to align its long axis perpendicularly to the axis of the maximum compressive stress. This alignment is energetically favored because of a volume change associated with the phase transformation. When the transformation to phase B is sufficiently fast, the lens-shape nuclei cannot support the applied stress, so its tips are subject to compressive stress, just as a crack is. Because of this additional compressive stress, phase transformation further progresses at the tips. The lenses of the high-pressure phase (phase B) thus propagate in a cracklike manner by interacting with each other, and they are eventually connected to form a fault. Shear motions on a fault may be possible under high pressures if the newly formed spinel phase has very fine grain size. The propagation of this lens-shaped high-pressure phase may also be facilitated by the latent heat release associated with phase transformation. The olivine-spinel transition is exothermic, so heat is generated during transformation. The resultant increase in temperature may enhance the instability.

So, in short, this model for deep earthquakes implies that (1) the olivine to wadsleyite or ringwoodite transformation occurs at a much greater depth than the depth of equilibrium transformations and (2) these transformations lead to runaway instability. The nonequilibrium transformation implies that there must be metastable olivine in the central cold portion of subducting slabs. These regions should have anomalously lower seismic wave velocities than the surrounding regions. Therefore seismological observations would identify this metastable olivine wedge. Takashi Iidaka and Daisuke Suetsugu, then at the University of Tokyo, reported that they indeed found this metastable olivine wedge in the subducting slab beneath Japan (1992), and this was thought to provide a strong support for the transformation faulting model. However, later detailed analyses showed that the detection of hypothetical metastable olivine wedge is exceedingly difficult, and even in the coldest slab at the Tonga subduction zone, where the metastable olivine wedge would be best developed, the currently available data do not provide positive support for the presence of a metastable olivine wedge (Koper et al. 1998). Thus, the presence of metastable olivine in subducting slabs has not yet been confirmed. The presence of a well-developed metastable olivine wedge is not obvious from the mineral physics point of view either. Based on the analysis of existing experimental data on the kinetics of phase transformations, Mosenfelder and others at Bayreuth Geoinstitut also discussed that the presence of a metastable olivine wedge to ~ 650 km depth is not conclusive (Mosenfelder et al. 2001). If a significant amount of water is present in subducting slabs, as some scientists suggest (e.g., Seno and Yamanaka 1996; Peacock 2001), then the kinetics of phase transformation will be significantly enhanced (Kubo et al. 1998), and therefore the presence of an appreciable metastable olivine wedge would be highly unlikely.

A more serious difficulty is the observed large width of faults associated with deep earthquakes. The transformation-faulting model assumes that the olivine-ringwoodite (or wadsleyite) transformation directly leads to deep earthquakes. Therefore, the faults of deep earthquakes must be nucleated in regions where the olivine-ringwoodite transformation (and other similar transitions as well) takes place. According to the estimated thermal structure of a subducting plate for the depth range of 500–650 km, only the central portion of the plate is cold enough for the metastable olivine to ringwoodite (or wadsleyite) transformation to occur, and the width of the cold part is estimated to be ~ 10–20 km (fig. 4-3). Consequently, the fault width for deep earthquakes should be similarly narrow. However, many large deep earthquakes, especially the Bolivian and Fijian Earthquakes of 1994, have a fault width of as much as 40–50 km (fig. 5-4). Therefore, this observation is inconsistent with the transformation faulting model. One way to get around this problem would be to assume that although faulting *nucleates* in the cold central part of subducted slabs, the

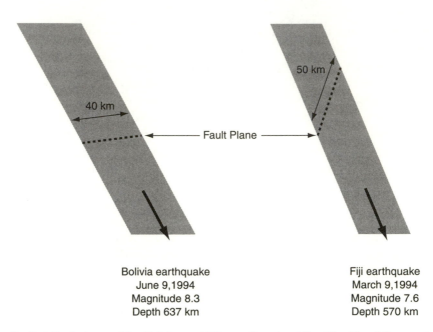

Bolivia earthquake
June 9,1994
Magnitude 8.3
Depth 637 km

Fiji earthquake
March 9,1994
Magnitude 7.6
Depth 570 km

Fig. 5-4. Fault planes of the Bolivian and Fijian earthquakes (after Kikuchi and Kanamori 1994; Wiens et al. 1994). Both planes are 40–50 km wide, far exceeding the prediction of the transformation faulting model (10–20 km).

fault plane estimated from seismological observations corresponds to the plane through which the fracture *propagates*. Fracture propagation would be possible even in regions where the phase transformation does not occur, so in this case, such an observation does not contradict the transformation faulting model. Such an argument cannot be defended, however. According to Doug Wiens and his colleagues at Washington University at St. Louis (Wiens et al. 1994), the aftershocks of the Fijian earthquake occurred in regions of ~ 50–km width. This means that earthquakes were *nucleated* in this broad region.

Yet another way to defend this model would be to argue that subducted slabs in these regions are much thicker than a simple thermal model would predict, and therefore the metastable olivine wedge is unusually thick there. Such a notion does not appear reasonable, however, because if the metastable olivine wedge is as thick as ~ 50 km at a ~ 650 km depth, it would cause a more serious problem; strong buoyancy is expected from the thick olivine layer, and it is difficult to understand why the plate is subducting there.

Furthermore, although there have been a number of pioneering studies by Steve Kirby and Harry Green, the processes of transformation faulting are not well understood. First, the conditions under which unstable deformation and the resulting faulting could occur are not known quantitatively. Therefore, even though faulting is observed under laboratory conditions, it is not clear if faulting would occur when transformations occur under geological conditions (likely at much slower rates). Second, the deformation experiments under high pressures are still highly challenging, and some of the follow-up studies under conditions similar to those simulated by Green and his colleagues (1990) did not reproduce the faulting instability reported earlier (Dupas-Bruzek et al. 1998).

Wiens and Snider (2001) reported "repeated deep earthquakes"—namely, two earthquakes that occurred at a nearly identical place within a short interval—and proposed that this observation is inconsistent with the transformational-faulting model; according to this model, no earthquakes will occur after a region has already transformed to a new phase. Because of these difficulties, I consider the transformation-faulting model to be less viable than previously thought.

Charlie Meade and Raymond Jeanloz, at the University of California at Berkeley, proposed another model. They found that some hydrous minerals, such as serpentine, become amorphous (i.e., disordered like glass) under high pressure and that these amorphous materials often show unstable deformation leading to faulting (under high pressures). They proposed that this unstable deformation caused by amorphization of hydrous minerals might be the cause of deep earthquakes (Meade and Jeanloz 1991). Several people have proposed that intermediate earthquakes may occur due to instabilities related to a dehydration reaction (Green and Houston 1995; Kirby, Engdahl, and Denlinger 1996a). For this model to explain deep (or intermediate-depth) earthquakes, subducted slabs must contain hydrous minerals in their central portions. Paul Silver suggested that subducted slabs could be hydrated due to normal faulting near trenches; in this case, he argued, the large width of deep earthquake faults can be explained (Silver et al. 1995). He and his colleagues proposed that deep earthquakes occur along the preexisting faults, where some hydrous minerals could be present. Paul Silver says, "Old faults never die." If this model is correct, we should expect a good correlation between the activity of deep earthquakes and the activity of shallow earthquakes before subduction. In addition, the fault plane orientation of deep earthquakes has to be determined by the orientation of shallow faults. Such a correlation is not very clearly observed, however. We also do not know how amorphization can cause instability, and Tetsuo Irifune, at Ehime University, and his colleagues showed that the amorphization of serpentine does not occur under deep slab conditions (Irifune et al. 1996). A perhaps more fundamental difficulty is that the volume change associated with dehydration reactions is usually negative in the deep mantle, and in these cases the instability of deformation does not occur (Wong, Ko, and Olgaard 1997). For these reasons, I consider that this model is, at best, speculative.

There is a completely different model for faulting. David Griggs, at the University of California at Los Angeles, the father of rock deformation studies, proposed that thermal runaway instability might be responsible for deep earthquakes (Griggs and Handin 1960; Griggs and Baker 1969) under high pressure, in which the instability of deformation due to thermal feedback causes melting and deep earthquakes. Masaki Ogawa, at the University of Tokyo, and Bruce Hobbs and Alison Ord, then at Commonwealth Scientific and Industrial Research Organization in Melbourne, made some detailed analyses (Ogawa 1987; Hobbs and Ord 1988). The essence of this model is as follows. The ductile deformation of minerals is enhanced by heat, and ductile deformation results in heat generation. Therefore, if thermal diffusion is slow compared to the rate of heat generation (by deformation), deformation is enhanced by the heat produced by the deformation itself. This is a self-accelerating process (positive feedback) and therefore would lead to runaway instability, and eventually to the melting of a deforming material along a localized region. This process, called *adiabatic instability* or *thermal runaway instability*, and has been well known in material science for a long time. The study of rocks from northwestern Italy has shown that this process can actually cause the melting of mantle materials and resultant rapid fault motion (Obata and Karato 1995). Kanamori, Anderson, and Heaton (1998) also discussed the possibility of melting on a fault plane associated with deep earthquakes.

The instability through this mechanism is closely related to the rheological properties of materials, and therefore the better understanding of flow laws of deep-mantle materials is critical to evaluate the plausibility of this model. Both Ogawa (1987) and Hobbs and Ord (1988) used the data of olivine rheology to evaluate this mechanism because this was the only data available at that time. Deep earthquakes do not occur in olivine-rich regions, so this is not satisfactory. Better data on rheological properties of deep mantle minerals are now available based on high-pressure studies on deformation and on transformation kinetics (see chap. 4).

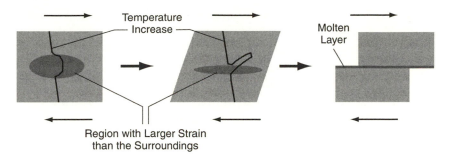

Temperature Increase

Molten Layer

Region with Larger Strain than the Surroundings

Fig. 5-5. A schematic drawing showing the processes of thermal runaway. When a certain region is deformed more than the surroundings, then more heat is generated in that region. This leads to a reduction of viscosity in that region, which further enhances deformation which results in more heat generation. This process potentially leads to thermal runaway (causing melting and earthquakes), if heat generated is not dissipated by thermal conduction and the degree of feedback is large.

Using a rheological model based on new experimental and theoretical studies, Karato, Riedel, and Yuen (2001) showed that this thermal runaway instability is a plausible model for deep earthquakes.

Let us discuss this model in some detail. Consider a deforming material (fig. 5-5). Let us assume that a small region of the material deforms slightly more than other parts. If this region becomes more easily deformable, then deformation in this region will further progress. Therefore, as far as external energy is supplied, deformation will continue to proceed, eventually resulting in melting. The condition for this instability is therefore equivalent to the condition for strain softening and therefore,

$$\frac{d\sigma}{d\varepsilon} = \frac{\partial\sigma}{\partial\varepsilon} + \frac{\partial\sigma}{\partial T}\frac{\partial T}{\partial\varepsilon} < 0, \tag{5-3}$$

where the first term denotes the explicit strain dependence of strength and the second term denotes the strain dependence of strength through temperature variation. The first term corresponds to *work-hardening* and is usually positive. However, the second term, corresponding to the effects of shear heating, is usually negative, because strength decreases with temperature ($\frac{\partial\sigma}{\partial T} < 0$; see equation [5-2]]) and strain causes heat ($\frac{\partial T}{\partial\varepsilon} > 0$). Consequently, if the softening due to shear heating is sufficiently strong, then the condition for instability (5-3) will be met and the faulting would occur.

Two issues are critical for this instability. First, in order to convert strain energy to heat, deformation must occur faster than heat diffusion. Otherwise, the heat generated by deformation will diffuse away and the temperature would not go up significantly. Second, the feedback from temperature increase to deformation must be strong. As shown in box 5-2, this feedback is stronger at lower temperatures. Therefore this *thermal runaway instability* occurs when deformation is fast at relatively low temperatures. It is not easy to meet these two conditions because the deformation rate is usually low at low temperatures. Indeed, earthquakes do not occur everywhere; they occur only in low-temperature regions under high stresses. In more detail, we can see that this mechanism is possible at a certain temperature range. If the temperature is too low, the deformation rate is so slow that nothing happens. If the temperature is too high, thermal feedback is not efficient, resulting in homogeneous deformation. Thermal instability takes place only at intermediate temperatures. These conditions are satisfied slightly off the central part of a subducting slab. Its temperature is relatively low, but the region is associated with high stress, so that fast deformation is possible.

Using this type of argument, Hobbs and Ord (1988) proposed that intermediate-depth and deep earthquakes might be caused by the thermal instability of deformation. Karato, Riedel, and Yuen (2001) applied their model with more realistic rheology and showed that indeed the conditions for thermal runaway instability are met in the deep portions of subducting slabs and that this instability is enhanced by phase transformations (fig. 5-6). Phase transformations are critical for two reasons. First, phase transformations lead to small grain size in the cold regions of slabs, causing dramatic reductions in strength (see chap. 4) which makes it possible for a cold slab to deform at a fast strain rate. Second, exothermic phase transformations such as the olivine to ringwoodite transformation generate heat, and hence enhance thermal runaway instability. In short, in this slab rheology model, the fast deformation of a slab as a whole becomes possible in the deeper portions due to weakening caused by phase transformations. Within a deforming slab (by bending), central regions are so weak that not much heat is generated. The outer region is very hot and weak, and again not much heat is generated by deformation. In the intermediate regions, the rate of deformation is fast yet stress is reasonably high. It is in this region where the conditions for instability would be met (see fig. 5-6). The width of this region can be estimated from the rheology model and is shown to be ~ 40–60 km which is in excellent agreement with the width of deep-earthquake faults.

Box 5-2. Thermal Runaway Instability

Thermal runaway instability is the mechanism for faulting by thermal instability. Suppose that a part of a rock somehow deforms more (faster) than other parts. In this part, more heat is generated due to deformation, and its temperature increases. If deformation is so fast that heat accumulates before diffusing into other parts, deformation becomes concentrated in this part. Eventually, this part melts, and abrupt faulting takes place. For this instability to happen, not only must deformation be fast, but the temperature also must be low enough. To efficiently enhance deformation by temperature increase, the original temperature cannot be too high. Fast deformation at low temperatures usually takes place with difficulty, but it is exactly what we expect to happen in a subducting plate. Thus, this model is the most promising mechanism for deep earthquakes. According to the plate rheology model shown in figure 4-7, this instability is expected not in the central part of a plate, but in its surrounding region, which is characterized by high strength and low temperature. The width of the region is calculated to be 40–60 km; this distance is consistent with the seismic observations of the Bolivian and Fijian earthquakes (fig. 5-5).

To evaluate this condition, we use the generalized form of equation (5-2),

$$\sigma = C\varepsilon^m \dot{\varepsilon}^{\frac{1}{n}} exp\left[\frac{H_n^*}{nRT}\right]. \tag{B5-2-1}$$

Using this equation, the condition in equation (5-3) may be expressed as

$$\frac{m\sigma}{\varepsilon} - \frac{\sigma H_n^*}{nRT^2}\left(\frac{\partial T}{\partial \varepsilon}\right) < 0. \tag{B5-2-2}$$

Next consider the term $(\partial T/\partial \varepsilon)$, which denotes how deformation increases temperature. Work done by stress is $\sigma\varepsilon$, so we can write

$$\frac{\partial T}{\partial \varepsilon} = \frac{K\varepsilon\sigma}{\rho C_p}, \tag{B5-2-3}$$

where C_p is specific heat. The parameter K denotes the conversion efficiency of external work to heat. When all work is converted to heat, $K = 1$; when no work is converted to heat, $K = 0$. This con-

(continued)

version efficiency is determined by the rate of deformation. When deformation is faster than thermal diffusion, work is efficiently converted to heat; when deformation is slow, generated heat escapes by diffusion, so temperature is not increased very much. The time-scale of thermal diffusion is given by L^2/κ (L is the length scale, and κ is thermal diffusivity), so this efficiency can be approximately expressed as

$$K = 1 - exp\left[-\frac{\dot{\varepsilon}L^2}{\kappa}\right]. \qquad \text{(B5-2-4)}$$

Thus, the conditions for deformation instability are

$$\frac{\dot{\varepsilon}L^2}{\kappa} > 1 \qquad \text{(B5-2-5)}$$

and

$$\frac{\sigma\varepsilon H_\eta^*}{nRT^2} > \frac{m}{\varepsilon}. \qquad \text{(B5-2-6)}$$

As can be seen from equation (B5-2-6), this mechanism is effective at relatively low temperatures. This is because the rate of softening due to temperature increase is greater for lower temperatures. On the other hand, to satisfy equation (B5-2-5), the deformation rate must be sufficiently high.

5-5. SUMMARY AND OUTLOOK

As you can see in the above discussion, I consider that the thermal runaway instability is the most plausible mechanism for deep (and also intermediate-depth) earthquakes for the following reasons: (1) this mechanism has a clear physical basis, its theory is well established, and instability due to this mechanism has been demonstrated experimentally for metals; (2) according to a new model of slab rheology, the conditions for thermal instability are likely satisfied in the deep portions of subducting plate; and (3) there is field evidence for the melting of mantle materials along a fault plane caused by this instability. Furthermore, the problem of wide faults (40–60 km wide), which cannot be explained by the transformational-faulting model, can be naturally explained by this thermal instability model. A small number of aftershock activities may also be ex-

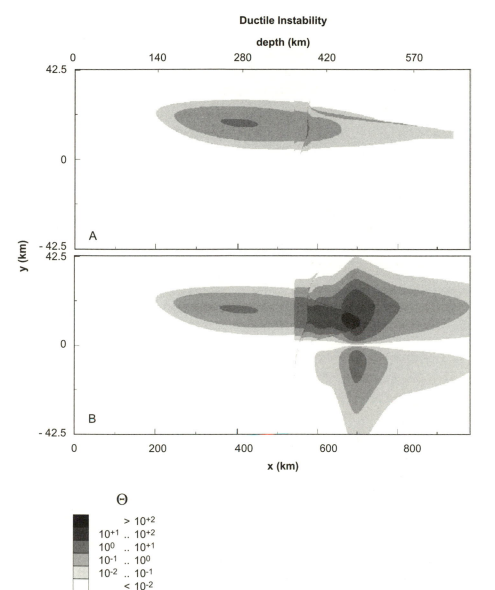

Ductile Instability

Fig. 5-6. Conditions for thermal runaway instability in a subducting slab (Karato et al. 2001). Values of a parameter Θ are plotted in a cross-section of a slab $\left(\Theta = \dfrac{\sigma \varepsilon H_2^* \kappa}{nRT^2 \rho C_p} \right)$. This parameter characterizes the condition for thermal runaway instability. When it is large (>1), thermal runaway occurs; thermal runaway does not occur when it is small (<<1) (for details see box 5-2). (a) Without the effects of phase transformation and (b) with the effects of phase transformation. The results show that phase transformations enhance thermal runaway instability. In this calculation, only the deformation due to the bending moment is considered.

plained as follows. Because of melting along a fault plane, stress is almost completely released (a part of stress is also released by viscous flow). Thus, little stress would be left to generate aftershocks.

This model can also explain the depth distribution of seismic activities. The transformational-faulting model need to have different models for intermediate-depth and deep earthquakes. The thermal instability model can explain both types of earthquakes in a unified fashion: as Hobbs and Ord (1988) showed, intermediate-depth earthquakes can also be explained using thermal instability.

The low efficiency of deep earthquakes and their slow rupture propagation are also favorable to this model. In this model, earthquakes occur not by perfectly brittle fracture but as the result of instability associated with viscous flow, so a part of the initial stress is relaxed by viscous flow. Thus, the efficiency of an earthquake should be generally low. In a visco-elastic fracture, the effective elasticity of a material decreases (chap. 2), so rupture propagation is expected to be slow.

How do we, then, test this model? An experimental test would be desirable, but this is not so easy because it is difficult to achieve the condition for adiabatic deformation (equation [B5-2-5]) in a laboratory. To see this point, consider a typical sample size of 3 mm. Since $\kappa = 10^{-6}$ m^2/s for mantle materials, the condition for adiabatic deformation will be met for strain rates higher than $\dot{\varepsilon} = 10^{-1}$ s^{-1}. This is much greater than the strain rate used in the usual deformation experiments ($10^{-6} - 10^{-3}$ s^{-1}). In fact, the thermal instability model is a plausible model for Earth's mantle because a deforming region is so large that thermal diffusion is not effective. With high-speed deformation experiments with large differential stresses, we may be able to detect thermal instability in the lab, but this is not a trivial issue, particularly under high pressures.

Although an experimental test would be wonderful if we could do it, numerical experiments using realistic mineral properties are also an effective approach. It is important to investigate what kind of characteristics of deep earthquakes this instability model would predict. In particular, it is essential to see how well this model can explain the efficiency of earthquakes, the characteristics of aftershocks, and the rate of rupture propagation.

Finally, I should also emphasize that the instability caused by dehydration reactions is a possible mechanism for deep and intermediate-depth earthquakes. Two issues must be further investigated for this hypothesis. First, the mechanisms for the hydration of central portions of subducting slabs are unknown and must be investigated quantitatively; the high water content in slab interiors must be demonstrated through some geophysical observations. Second, the nature of the instabilities associated with this mechanism should be investigated in more detail, particularly in relation to the known characteristics of deep (and intermediate-depth) earthquakes. In particular, if instability were to occur associated with reactions where the volume shrink occurs, the physical mechanisms must be explained.

More than seventy years after the discovery of deep earthquakes by Kiyoo Wadati, we now have much better ideas, due to the efforts of a number of scientists in different areas, about materials properties and the processes of faulting in the deep mantle. Much progress has been made, particularly in the areas of materials properties, including the nature of phase transformations and mechanical (rheological) properties under deep-mantle conditions. However, the major obstacle in identifying the mechanisms of deep earthquakes is the difficulty in conducting well-controlled deformation experiments under deep-mantle conditions. Therefore, we must conclude that the origin of deep earthquakes is still enigmatic. New techniques for deformation experiments under high pressures have recently been developed, however (see chap. 3), and we expect major progress in the near future that will help us to solve this long-standing mystery.

SIX • THE CORE: STRUCTURE, EVOLUTION, DYNAMICS, AND PLANETARY PERSPECTIVES

6-1. THE STRUCTURE OF THE CORE AND ITS DYNAMICS

Let us now look into the deepest portion of Earth—namely the metallic core. It is Earth's (and other planets') core where the magnetic field is believed to be generated. In addition, the structure and evolution of the core are closely related to the evolution of terrestrial planets, so we can acquire key information about planetary evolution from observations regarding the core. The study of the core and the geomagnetic field has been isolated from other areas of Earth sciences. However, owing to recent discoveries in seismology, high-pressure studies on core material, progress in dynamo theory, and an improved understanding of the cores and magnetic fields of other planets, the study of the core has become highly interdisciplinary beyond our anticipation.

6-1-1. The Composition and Structure of Earth's Core

Earth's core consists of the liquid outer core and the solid inner core (fig. 6-1). The core is an almost iron-nickel alloy (we simply call it iron, below, because the amount of nickel is presumably small, a few percent), though the composition of the inner and outer core are slightly different. The density of the outer core is significantly smaller than the density of iron estimated from high-pressure experiments. Consequently, the outer core is considered to contain some light elements, but the inner core is closer to pure iron.

Light elements that have so far been considered as impurities in the core include sulfur, oxygen, hydrogen, silicon, and carbon (Poirier 1994). All of these elements are abundant in the solar system. However, each of them

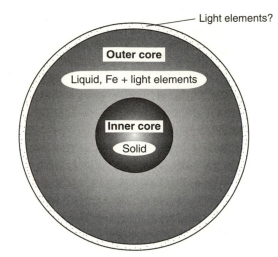

Light elements?

Outer core

Liquid, Fe + light elements

Inner core

Solid

Fig. 6-1. Structure of the Earth's core. The outer core is liquid, and the inner core is solid. At the surface of the inner core, there may be a partially molten layer (the F layer).

has different solubility in iron, which depends on thermodynamic conditions such as pressure, temperature, and oxygen fugacity. Among these elements, only sulfur can be dissolved into iron in substantial amounts under relatively low temperatures and pressures. The others can be dissolved into iron under high temperatures and pressures, but their solubility under low temperature and pressure is limited. In fact, iron meteorites, which are considered to have come from small planetary bodies (a size of ~ 1,000 km or less), must have reached chemical equilibrium under relatively low pressures (< 5 GPa), since a large amount of sulfur is dissolved but the other elements are scarce. Thus, if chemical equilibrium between the core and the mantle occurs under high temperature and pressure corresponding to the present core-mantle boundary, most of the above elements can be largely dissolved. However, if the chemical equilibrium was established under lower temperature and pressure, sulfur is the most likely candidate.

The presence of a significant amount of light elements in the outer core (but not in the inner core) is a key to the understanding of the evolution and dynamics of the core. Molten metals, including iron, can dissolve large amounts of impurities, particularly at high pressures and temperatures, but the solubility of impurities is much less for the solids. Therefore, the inferred high impurity content in the liquid outer core but the small amount in the solid inner core can be explained by assuming that the inner core has been formed as a result of solidification from the originally (totally) molten core. Upon solidification, larger amounts of impurities go to outer core, and the inner core, with relatively pure iron, grows. This solidification of the outer core produces latent heat, and light elements become concentrated in the outer core; much energy is produced at the inner-outer core boundary by the growth of the inner core. Stanislav Braginski, then at Moscow, and now at the University of California at Los Angeles, was the first to point out that the growth of the inner core provides an important energy source for the geodynamo (1963).

Furthermore, if the inner core has been formed from the outer core by solidification, the concentration of impurities in the outer core must have increased with time. As a result, the concentration of impurities may have eventually exceeded the solubility limit in the outer core. Then, impurities (i.e., light elements or compounds containing light elements) would have been precipitated as separate phases and would have floated to the upper part of the outer core as "sediments." Based on recent seismic and geodetic observations, Bruce Buffett, at the University of British Columbia, inferred the existence of a layer with finite rigidity at the topmost part of the outer core (Buffett, Garnero, and Jeanloz 2000). Some parts of the D″ layer may contain materials accumulated from the core below, in addition to the materials accumulated from the mantle above. The ultra-low-velocity region in the D″ layer may be a patch of this garbage-rich region coming originally from the outer core.

What is the nature of the transition from outer to inner core? There have been some speculations about the gradual transition from the liquid outer core to the solid inner core. According to the theory of solidification, when some conditions are met, the growing solid assumes an irregular morphology known as *dendrite*. David Fearn and his coworkers argued that dendritic growth of inner core occurs, and that the inner-outer core transition is so gradual that this *mushy layer* may extend to the center of the inner core (Fearn, Loper, and Roberts 1981). However, Ikuro Sumita and his colleagues at the University of Tokyo (Sumita et al. 1996) showed that when the compaction due to gravity is considered, such a layer will not survive if the permeability is high enough.

The inner core is a small spherical region (with a radius of 1,220 km), which occupies only ~ 4% of the core and ~ 0.7% of whole Earth. However, its important role on dynamo action has recently been recognized. The inner core is made of mostly solid iron. Its density and elastic constants are consistent with those of pure iron (with a small amount of nickel). Although some people have speculated on the presence of large porosity (i.e., partial melting), both the seismic wave velocities and attenuation are consistent with the solid state without melt. The most surprising finding about the inner core is its elastic anisotropy and possible superrotation (rotation faster than the mantle). These issues are closely related to the dynamics of the inner core and will be discussed in some detail in section 6-3.

6-1-2. The Evolution of the Core

When and how was Earth's core formed? This question has attracted much attention because the processes of core formation affect the chemistry of both the core and mantle and the history of the magnetic field. The age of the formation of the core can be inferred from geochemical measurements using geochemical characteristics of some elements (isotopes). One of the isotopes of hafnium, ^{182}Hf, is a radioactive isotope that changes to ^{182}W (an isotope of tungsten) through beta-decay and has a half-life of 9 million years. Since W (tungsten) is a siderophile (iron-loving) element, ^{182}W goes to iron and is removed from silicates. Therefore, if the core formation occurred early, when a still significant amount of ^{182}Hf was there, then ^{182}Hf would have created ^{182}W in the mantle, which would be detected as an excess of W (more than the abundance in the chondrite, which has not undergone iron separation). Using this technique, Der-Chuen Lee and Alex Halliday, then at the University of Michigan, determined the age of formation of Earth's core (Lee and Halliday 1995). They showed that the separation of iron and silicates in Earth is likely to have occurred ~ 62 ± 10 million years after the time of formation of iron meteorites. Thus, core formation occurred in the very early stages of Earth evolution, perhaps almost contemporaneous with the formation of Earth, although it took some time after the formation of small planetary bodies.

The processes of core formation are more controversial because the observational constraints are indirect. However, the processes of core formation have an important influence on the composition of both core and mantle. If iron and the mantle have been separated from a mixture maintaining chemical equilibrium, then a significant fraction of water (hydrogen) must go into the core (Fukai 1984; Okuchi 1997). Similarly, the abundance of light elements in the core and the abundance of siderophile (iron-loving) elements in the mantle will be controlled by the conditions at which the final chemical equilibrium was attained.

The most important constraint for the formation process of the core is

Box 6-1. Siderophile Elements and Core Formation

The equilibrium concentration of elements in a given multiphase material is determined by the affinity of that element to the existing phase. Some elements prefer iron and others prefer rocks. The former are referred to as *siderophile* (iron-loving) *elements* and the latter as *lithosphile* (rock-loving) *elements*. Whether an element belongs to either the siderophile or the lithophile elements depends on the nature of element (the size that depends also on the charged state, the nature of the state of the electrons in its outer shells, etc.). Ni (nickel) and Co (cobalt), which belong to the same class of metallic elements in the periodic table (a group called "transition metals"), are typical siderophile elements. Magnesium, on the other hand, usually assumes Mg^{2+} and is easy to accommodate in silicates rather than in metals (a small amount of Mg could go into iron under extremely low oxygen fugacity conditions). When iron and silicates coexist at high temperature over a long time, the abundance of these elements in each phase is controlled by chemical equilibrium; most of siderophile elements will go to iron and not to silicates. Assuming chemical equilibrium and knowing the total abundance of the element and the abundance of iron and silicates, we can calculate how much of the siderophile elements must be in silicates if silicates are in chemical equilibrium with iron (at a certain pressure and temperature). Ringwood did such a calculation using experimental data at moderate pressures and temperatures, and concluded that the amount of Ni and Co in the upper-mantle rocks is too high if the upper mantle silicates were in equilibrium with iron during core formation. These observations can be interpreted as either disequilbirum in the core formation or as chemical equilibriums at the different pressures and temperatures that control the chemical composition of the mantle.

the abundance of siderophile elements in the mantle (box 6-1). The abundance of these elements tells us how iron interacted with silicates during the core formation processes. Ted Ringwood was the first to note that the chemical composition of Earth's mantle is not in equilibrium with the core (1966). He noted, among other things, that the mantle has too much ferric iron (Fe^{3+}) and too much Ni (and other siderophile elements; see fig. 6-2) for chemical equilibrium with iron and suggested that the core was

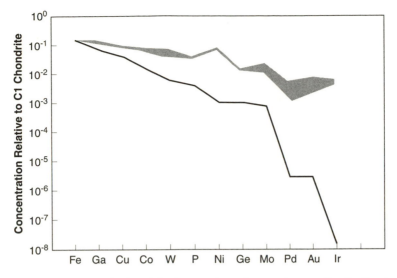

Fig. 6-2. The abundance of siderophile elements in Earth's mantle normalized by the abundance in the carbonaceous (C1) chondrite (hatched region) (from Murthy and Karato 1997). The observed abundance of siderophile elements is much higher than that expected from the experimental results at low-*T* (and *P*) (shown by a solid line).

formed in a process in which iron was out of equilibrium with silicates, and the core and the mantle have since remained out of chemical equilibrium. Several different models have been proposed to explain this observation. The first and the most radical model is the one proposed by Karl Turekian and Sydney Clark, at Yale University (1969), who proposed that during the formation process of planets, the core was first formed, then silicate mantle was accumulated on top of the preexisting core. A possible reason for this is that iron could have been condensed earlier than silicates in the primitive solar nebula, and that iron accumulated quickly to form the core, then silicates accumulated on top of it. This model would easily explain the disequilibrium between the mantle and the core. However, based on detailed thermodynamic calculations, we now know that silicates and iron likely have condensed at similar stages to form a fine-scale mixture (Grossman 1972). According to Larry Grossman's calculation, Ca- and Al-rich refractory oxides were first formed in the cooling solar nebula, and iron and most of silicate minerals were formed at nearly the same temperatures in a later stage. This prediction is supported by the observations of primitive meteorites such as the carbonaceous chondrite (see chap. 1) which sometimes show "white inclusions" containing Ca-

and Al-rich refractory oxides and a fine-scale mixture of metallic iron and other silicate minerals. Therefore, a likely process of planetary formation is that small fragments of a nearly homogenous mixture were first formed and then, through their collisions, progressively larger bodies were formed. They include putative parental bodies for several "differentiated" meteorites, such as achondrite (a type of meteorites that shows evidence of metamorphism) and iron meteorites. These meteorites are considered to have been formed in small planetary bodies (a few 100 km to ~ 1,000 km size) through some chemical differentiation processes: iron meteorites, for example, are considered to be from the cores of these small planetary bodies, which were later fragmented by mutual collisions and then distributed in the asteroid belt. The major process of planetary formation therefore includes the collision of small planetary bodies with a range of sizes. The final stage of formation may have included collision of large bodies (> 1,000 km), which could have shattered a proto-Earth to form the Moon (e.g., Wetherill 1990). In short, materials from which Earth was formed are believed to be a mixture of iron and silicates, although the scale of mixing depends on the size of these bodies.

Therefore, most of the models of core formation address the question of how iron and silicates could have been separated, keeping an abundance of siderophile elements at the observed level. Rama Murthy, at the University of Minnesota (1991), suggested that the partitioning of elements between iron and silicates is likely to depend on temperature, and at very high temperatures, the difference between siderophile (iron-loving) and lithophile (rock-loving) elements will be diminished. He argued that if the separation of iron and silicates occurred at very high temperatures (say ~ 3,000–4,000 K) as some of the models of planetary formation suggest, then the observed abundance of siderophile elements could be explained. This paper had a strong impact to the geochemical community, where people had been using the experimental data on element partitioning obtained at modest conditions to discuss the formation of Earth's core, where extreme conditions are likely to have been involved. This paper stimulated people to investigate element partitioning under extreme conditions. Jie Li and Carl Agee, then at Harvard, showed that the equilibrium distribution of siderophile elements such as Ni (nickel) and Co (cobalt) changes dramatically with pressure (and temperature) (2001). They showed that if the last chemical equilibrium occurred at ~ 45–55 GPa (~ 3,000 K), then the abundance of Ni and Co is consistent with the observations. Such a scenario is possible when the iron-silicate separation occurred in the magma ocean, whose depth corresponded to these pressure and temperatures (~ 1,200–1,450 km for the current size of Earth). One of the difficulties with these models is that the abundance of very highly siderophile elements such as Au (gold), Pd (palladium), and Ir (iridium) in mantle materials is nearly equal (relative to the chondrite values) (fig. 6-2): they do not show much element-to-element variation in abundance. If the abundance of these elements is controlled by thermodynamic equilibrium, it is more likely that the abundance (relative to chondrite) is different from one element to another.

Several scientists, including Kimura, Lewis, and Anders, at the University of Chicago (1974), proposed that the abundance of these siderophile elements is largely controlled by the primitive materials that were accreted on Earth in the later stage of Earth formation. These materials are called "later veneer," and are supposed to contain a large amount of siderophile elements. Another, conceptually similar model is the *incomplete core formation model* (Jones and Drake 1986). This model shows that the separation of iron from silicates was incomplete and that there was some "leftover" iron in the mantle, which was later dissolved into silicates. These leftover materials are assumed to contain a large amount of siderophile elements.

Dave Stevenson, at Caltech (1990), and Karato and Murthy (1997) investigated the physical and chemical processes of core formation. Two issues are critical. First, when a planet grows through the accretion of proto-planetary bodies, its temperature increases gradually with time because the amount of gravitational energy released upon the collision of these bodies increases with the size of a planet. Accordingly, in a growing planet, temperatures will be lower in the deeper portions (the opposite of the present Earth; see fig. 1-10). Consequently, the deep interior of a growing planet likely has a high viscosity, but the near surface regions of a relatively large planet will be hot and soft and could even be molten (magma ocean). Continuing collisions of proto-planetary bodies occur on a growing planet that has this type of rheological structure (fig. 6-3). Upon collisions, some of the bodies will be shattered and the molten material will form iron droplets, which will sink to the surface of a cold central portion, where viscosity is high. The size of droplets is controlled by surface tension and will be on the order of 1 cm or less. Therefore, these iron droplets will be in chemical equilibrium with the surrounding silicates during their descent. After the accumulation of a layer with some critical thickness, gravitational instability will occur, by which a large-scale overturn will bring a large mass of iron to the center of the proto-Earth. This process of iron separation likely occurs without chemical equilibrium with the surrounding materials. Therefore, in this model, the final chem-

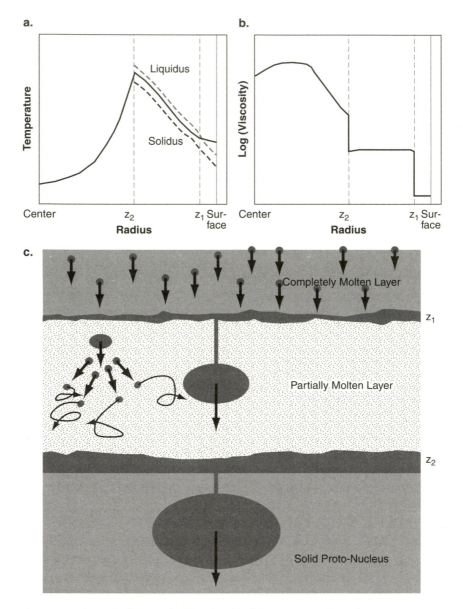

Fig. 6-3. A schematic diagram showing the plausible processes of core formation (Karato and Murthy 1997). (a) A temperature profile of a growing Earth. (b) A viscosity profile of a growing Earth. (c) Processes of iron-silicate separation in a growing Earth. The degree of chemical equilibrium during iron-silicate separation depends primarily on the size of the iron body. Separation of molten iron droplets through a completely molten magma ocean is likely to have occurred at chemical equilibrium because of the small size of the iron droplets, which is controlled by surface tension. However, the separation of a large iron body through a deep, partially molten magma ocean is likely to have occurred without chemical equilibrium. The separation of iron also could have occurred as a large body when large-size planetesimals collided to form Earth without chemical equilibrium.

ical equilibrium will be attained at the bottom of the magma ocean. Karato and Murthy (1997), motivated by the results of numerical modeling of giant impacts (Benz, Slattery, and Cameron 1989), further noted that collisions of large bodies may have injected the core materials of these large bodies directly into the proto-core of the growing Earth. If this occurred, a large fraction of core materials are directly from the cores of small planetary bodies because most of the mass of materials from which Earth was formed belongs to these large bodies. In this case, the chemical composition of Earth's core is largely determined by low-pressure (and moderate-temperature) chemical equilibrium and is similar to that of iron meteorites. The likely light element in the outer core in this scenario is sulfur. The chemical composition of the mantle, on the other hand, is controlled by the chemical equilibrium at the conditions within the magma ocean with a small amount of iron that did not go into the core. Murthy and Karato (1997) showed that the abundance of siderophile elements in the mantle can be explained if $\sim 0.3-0.5\%$ of iron of the core is in equilibrium with the mantle. So in terms of mass balance, this model is similar to incomplete core formation model, but is based on a different physical model of core formation.

Iron-silicate separation (i.e., core formation) is likely to have occurred almost contemporaneously with the formation of Earth. But the difference in densities and, hence, the composition between the outer and the inner core strongly suggest that the inner core has grown from molten iron through cooling. When did Earth's core start to freeze? Bruce Buffett and his colleagues studied the thermal history of the core (Buffett et al. 1992; Buffett 2000). Assuming both a constant heat flux from the core and no radiogenic heat generation in the core, they showed how the growth rate of inner core depends on the assumed heat flux. For example, if the heat flux from the core is 2×10^{12} W, it would take ~ 4.5 billion years for the inner core to grow to the present size. If the heat flux is 6×10^{12} W, it would take only ~ 2 billion years (fig. 6-4). How can we estimate the heat flux from the core? One way is to use the estimated heat transfer from plumes. Plume heat flux is estimated to be $\sim 5-10\%$ of the total heat flux, that is, $\sim 2-4 \times 10^{12}$ W (Sleep 1990). Heat from the core will also warm up the core-mantle boundary and can drive mantle convection. Though this heat flux can be calculated from the thermal gradient at the bottom of the mantle (the D" layer) and its thermal conductivity, this estimation is difficult to make. Considering various sources of errors, total heat flux from the core can be estimated to be $\sim 3-10 \times 10^{12}$ W.

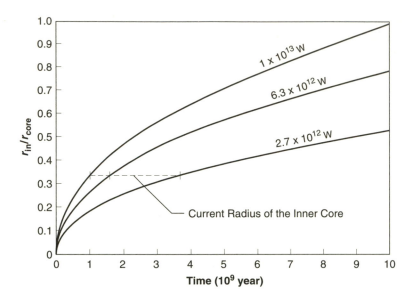

Fig. 6-4. The thermal history of the Earth's core (after Buffett et al. 1992). Values denote the heat flux from the core to the mantle. r_{in} is the radius of the inner core; r_{core}, the radius of the outer core.

A problem with this argument is that we do not know whether core heat flux has been constant. Heat from the core is transferred through the mantle, and heat transfer in the mantle depends on its temperature and water content. Thus, core heat flux is likely to have been larger in the past. The model by Buffett and his colleagues does not take this point into account. Considering the likely temporal evolution of mantle heat flux, the inner core may have formed in the very early stage of the Earth's history. Its growth rate may have been fast at the beginning and may have slowed down later. This is because, in the early history of the Earth, the interior was probably hotter and had a larger amount of water in the mantle (chap. 4).

The timing of the formation of the inner core is poorly constrained. However, if the existence of the inner core is necessary for the generation of a strong geomagnetic field (as I will discuss later), the formation time of the inner core can be inferred from the past intensity of the geomagnetic field. Based on this type of inference, Breuer and Spohn, in Germany (1995), estimated the formation time of the inner core as ~ 2.7 billion years ago, and they suggested that it might be the cause of the Archaean-Proterozoic boundary. Around this time, there were large changes in Earth's history such as increased volcanic activities and the resultant rapid growth of continental crust.

Such a model for the inner-outer core structure suggests that the structure of Earth's core is controlled by a delicate balance of its thermal evolution. In other words, if the heat source or cooling rate (or both) is somewhat different, then such a layered structure of the core would not be present. If a planet has too much heat or its cooling is sluggish, then the core would still be completely molten. If the planet has been cooled very efficiently, then the core would be completely solidified. In either case, one would not expect vigorous dynamo action, as I will discuss in the next section. For example, there is evidence that Mars once had a strong magnetic field (Connerney et al. 1999) in its very early stage (~ 4 billion year ago), but there is no strong magnetic field on Mars at present. A similar history of a magnetic field was proposed for the Moon by Keith Runcorn, then at Newcastle upon Tyne, in England (1983). In these planets, the inner-outer core structure may have existed in their early stage of evolution, but because of their small size, these planets have cooled rapidly, and now their cores are likely to be completely solid. Among the satellites of Jupiter, a recent *Galileo* mission showed that Ganymede has a strong magnetic field, Europa has a modest one, and Io and Callisto have very weak magnetic fields. Jerry Schubert, at the University of California at Los Angeles, suggested that such a variation in magnetic field strength might be due to the differences in the structure of their cores (1997): Io has large energy source due to tidal friction; therefore, its interior is hot and the core is likely to be totally molten. Ganymede's tidal heating is much less and its core, like Earth's, may have both liquid and solid portions. Callisto does not seem to have undergone much chemical differentiation and probably has no metallic core.

In view of this "comparative planetology" perspective, the Earth-Venus contrast in terms of their magnetic fields (as well as other activities) is striking and puzzling. Earth and Venus have nearly identical sizes and densities, and therefore many people expect that these planets have similar structure and dynamics. Contrary to this expectation, Venus shows a very weak magnetic field (and no evidence for plate tectonics). Dave Stevenson and his colleagues (Stevenson, Spohn, and Schubert 1983) proposed that because Venus is slightly smaller than Earth, the pressure in deep Venus is lower, and therefore the melting temperature of core materials will be lower. Assuming that Venus and Earth have similar temperature-depth profiles, they suggested that the core of the Venus remains totally molten and, hence, no strong magnetic field is generated. Another possibility is that the cooling of core of Venus might be more sluggish than that of Earth's core and that the Venusian core may be hotter than Earth's core. The rate of the cooling of the core is controlled by the rate of the cooling of the mantle. Based on the measurements of the composition of surface materials, Venus is believed to have a much smaller water content than Earth. Water significantly affects the viscosity of mantle materials (chap. 2), which controls the rate of cooling by convection. Therefore, Venus's interior is likely to be hotter than Earth's and its core may still be totally molten. Another striking difference between Earth and Venus—the absence of plate tectonics on Venus—may also be due to this absence of water (chap. 2).

6-2. THE DYNAMICS OF THE OUTER CORE: THE DYNAMO

The study of the geomagnetic field was pioneered by the German mathematician Friedlich Gauss, at Göttingen in the nineteenth century. Gauss clearly showed mathematically that the geomagnetic field is dominated by a dipole component, and that the cause of the magnetic field is internal, not from the outside of Earth. Later measurements also showed that the geomagnetic field has been varying with the human time-scale (several tens of years). The most obvious variation is the gradual migration of the pattern of the geomagnetic field toward the west, known as *westward drift*. In the mid-twentieth century, a method to estimate the ancient geomagnetic field from rocks was established, and an astonishing fact was revealed: the polarity of the magnetic field was repeatedly reversed in the past.

Because the origin of the geomagnetic field is in Earth's interior, it is quite certain that the magnetic field is generated by the dynamo process in the outer core. This is because (1) iron cannot have ferromagnetism like a permanent magnet under the temperature (and pressure) of the core, and (2) the geomagnetic field has secular variation, whose time-scale is too short to occur if it is generated in a solid portion. Therefore, it is believed that the magnetic field is generated by some effects of convection in the outer core, which is made of molten iron.

6-2-1. Energy Sources

If the dynamo process is generating the geomagnetic field, a heat source is necessary to maintain it. Otherwise, energy would be dissipated by electric currents creating a magnetic field. Though the amount of the required energy is difficult to estimate due to uncertainty about the intensity of the magnetic field and the time-scale of its variation, it is estimated as roughly 10^{11}–10^{12} W. To maintain the dynamo process, the energy source must produce energy exceeding the above value. Furthermore, convection in the outer core is necessary for the dynamo process, so the energy source also needs to be able to generate enough buoyancy. For example, in the case of thermal convection, in which buoyancy is created by thermal expansion, a conductive thermal gradient corresponding to a given heat source must exceed the adiabatic thermal gradient.

The decay of radioactive elements is an obvious candidate for the energy source. However, it is difficult to extract sufficient energy by only radioactive decay in the case of the core, because the core is composed of metals and its thermal conductivity is very high. This can be understood by calculating how much heat is transferred along the adiabatic temperature gradient in the core. The adiabatic temperature gradient in the core is ~ 0.7 K/km, and thermal conductivity is ~ 40 W/Km (this is about ten times larger than the thermal conductivity of rocks at the surface). From this calculation we estimate that the heat flux carried by the adiabatic thermal gradient is ~ 4×10^{12} W. Dividing this by the total mass of the core, one can calculate the amount of radiogenic heat production needed for the heat flux that exceeds this value. The calculation shows that the amount of radiogenic heat production must exceed 2×10^{-12} W/kg. This heat generation is almost the same as that in the mantle (the inner core must have heat production of more than 5×10^{-11} W/kg to exceed the adiabatic temperature gradient, and this value is greater than the radiogenic heat production of basalt). According to the study of iron meteorites, however, iron (and nickel) contains a much smaller quantity of radioactive elements compared to the amount in silicate rocks, so such high heat production is unlikely. Therefore, it is generally believed that the radiogenic heat production is not a likely source for the geodynamo.

Though cooling itself does not generate energy, if some gravitational energy released during the core (or Earth) formation is retained in the core, it can be gradually released to drive the fluid motion in the core. If the core is cooling at the rate of 100 K per billion years by releasing this ancient energy, ~ 5.7×10^{12} W of energy can be obtained. This is large enough to drive the dynamo, but the distribution of this energy is unlikely to be uniform. That is, although the secular cooling appears to be a plausible ultimate source of energy for the dynamo, the actual radial distribution of this released energy is likely to be concentrated at the inner-outer core boundary. When the inner core solidifies from the outer core, which contains abundant impurities, the concentration of impurities in the outer core increases because not many impurities enter the solid inner core. Therefore, the outer-core materials (liquid iron) near the inner-outer core boundary must be more enriched by impurities than are materials farther from the boundary. The impurities in the outer core are buoyant, so the bottom of the outer core has a relatively low density, leading to convection by gravitational instability. The density anomaly due to chemical heterogeneity is very large, and the gravitational energy associated with chemical composition does not diffuse as heat does. In addition, latent heat is also released with the solidification of the inner core. Therefore, the generation of (chemical and thermal) buoyancy at the inner-outer core boundary due to inner-core growth is the most important driving force of the dynamo process, and inner-core growth is thought to be required for the generation of a strong magnetic field.

6-2-2. Dynamo Theory

Although the detailed explanation of dynamo theory is beyond the scope of this book, I will explain some fundamental issues for completeness (for the explanation of dynamo theory, see Merrill, McElhinney, and McFadden 1998). The essence of the dynamo process is that when a fluid with high electric conductivity moves heterogeneously, this movement results in the distortion of magnetic field lines and, hence, generates a magnetic field. The generated magnetic field then exerts force on the fluid, affecting its motion in turn. The fundamental aspects of this interaction were revealed by Michael Faraday and James Maxwell, and the equations to describe this phenomenon—the equation of motion for viscous fluid and the equation of electromagnetic induction—had been established by the latter half of the nineteenth century. However, it is not easy to solve these equations directly, mainly because they are nonlinear. Therefore, the origin of the geomagnetic field has been considered one of the most difficult problems; Albert Einstein regarded this as one of the most important unsolved problems in physics.

The first person who studied this geodynamo process was Joseph Larmor, in Ireland, at the beginning of the twentieth century (1919). The first extensive studies of the dynamo were conducted during and soon after World War II by Walter Elsasser (1946), who earned his doctorate in physics and came to the United States from Germany, and by Edward Bullard (1949). Wartime studies of radar seemed to have sparked their interested in this problem in classical electromagnetism. However, it was difficult to simultaneously solve the equation of electromagnetic induction together with the equation of the motion of viscous fluid. In the early stage of the dynamo study, therefore, a simplified method was employed, in which the equation of electromagnetic induction was solved for a given (assumed) fluid motion. This is called the kinematic dynamo.

The calculation of the real dynamic dynamo, in which the equation of motion for rotating viscous fluid and the equation of electromagnetic induction are simultaneously solved, was conducted in the United States and Japan for the first time in the 1990s. In 1995, Gary Glatzmaier and Paul Roberts (1995a) were able to successfully simulate a magnetic field similar to the observed geomagnetic field. Using calculations by a powerful supercomputer at Los Alamos, they published a series of papers on the geodynamo, including the simulation of geomagnetic reversal (Glatzmaier and Roberts 1995b; Glatzmaier et al. 1999) and the prediction of inner core rotation (Glatzmaier and Roberts 1995a). At the same time, Jerry Bloxham and his colleagues at Harvard solved the problem by using different boundary conditions and obtained quite a different pattern of the magnetic field (Kuang and Bloxham 1997). In Japan, Kageyama and Sato have conducted similar studies (1997). Though the ability to reproduce the magnetic field similar to the actual geomagnetic field is a great step forward, the origin of the geomagnetic field has not been settled. Although these calculations are conducted under fairly realistic conditions, the approximation used for boundary conditions is still crude, and the pattern of the magnetic field varies significantly by changing boundary conditions. Furthermore, the effects of the inner core are not fully taken into account (I will explain this point later in this section in more detail). Important parameters such as the Rayleigh number (see chap. 1) are also still far from realistic values. The mechanism of the geomagnetic reversal is also not yet clearly understood.

To understand the dynamo theory, let us first examine the equation of electromagnetic induction, which can be derived from the Maxwell's equations and Ohm's law. The equation of electromagnetic induction is given by (see Merrill, McElhinney, and McFadden 1998)

$$\frac{\partial B}{\partial t} = \eta_M \nabla^2 B + \nabla \times (v \times B), \tag{6-1}$$

where B is the magnetic flux density, v is fluid velocity, and $\eta_M \equiv 1/\sigma\mu_0$ is the magnetic diffusion coefficient. If there is no fluid motion ($v = 0$), this equation reduces to the diffusion equation, $\frac{\partial B}{\partial t} = \eta_M \nabla^2 B$, so the magnetic field will disappear with the time-scale of $\tau \approx L^2/\eta_M = L^2\sigma\mu_0$ (σ is electric conductivity; μ_0 stands for magnetic permeability in a vacuum). Here, L denotes the typical scale of fluid motion (~ size of the core), and it can be shown that the magnetic field should disappear within about ten thousand years. Therefore, the fluid motion of the core is needed to maintain the magnetic field and also to control its temporal variations. For the magnetic field to exist, the second term (for electromagnetic induction) of equation (6-1) must be larger than the first term (for diffusion). Thus, the necessary condition for the presence of dynamo is

$$R_m \equiv \frac{\nabla \times (v \times B)}{\eta_M \nabla^2 B} \approx \frac{vL}{\eta_M} = vL\sigma\mu_o > 1, \tag{6-2}$$

where R_m is called the magnetic Reynolds number. Therefore, the dynamo process can take place when electric conductivity is high and fluid motion is fast. The dynamo process also occurs more easily in a larger object, such as a celestial body, including Earth (for this reason, the experimental study of the dynamo is difficult). If fluid velocity is assumed as ~ 10^{-4} m/s (this can be estimated from the westward drift of the magnetic field), we have $R_m \sim 150$ using the magnetic diffusion coefficient as assumed earlier. This suggests that the dynamo process is actually operating in Earth's core. Note also that equation (6-1) does not change by replacing the magnetic field from B to $-B$. Thus, two kinds of magnetic fields, with opposite polarities, are possible for the same pattern of fluid motion (v).

The fluid motion follows the Navier-Stokes equation. For simplicity, consider steady-state motion. In this case, the equation of motion is given by

$$2\rho\Omega \times v = -\nabla p + \eta\nabla^2 v + \rho'g + F, \tag{6-3}$$

where Ω is the rotation velocity vector of Earth and $2\rho\Omega \times v$ denotes the Coriolis force, η is viscosity, ρ' is density variation, including the effects of thermal expansion and chemical composition, g is the acceleration due to gravity, and F is the Lorentz force (force due to the magnetic field). The pattern of motion is determined by the balance of gravity ($\rho'g$), the pressure gradient, viscous friction ($\eta\nabla^2 v$), the Lorentz force, and the Coriolis force. The appearance of the Lorentz force and the Coriolis force in the equation of motion is a big difference from the case of mantle convection. The dynamo process occurs as the result of interaction through the Lorentz force between fluid motion and the electromagnetic field.

Let us compare the magnitude of viscous resistance, the Lorentz force, and the Coriolis force. The Taylor number defined by $T_A \equiv 2\rho\Omega L^2/\eta \approx |2\rho\Omega \times v|/|\eta\nabla^2 v|$ is a measure of the ratio of the Coriolis force and the viscous resistance. Assuming that the viscosity of the mantle is 10^{21} Pa·s, the viscosity of the inner core is 10^{16} Pa·s (this is estimated from interaction between the inner core and the mantle [Buffett 1997] and from the data of hexagonal metals that have the same crystal structure as the inner core [Karato 1999]), and the viscosity of the outer core is 10^{-3} Pa·s, the Taylor numbers of the mantle, the inner core, and the outer core are estimated to be ~ 10^{-7}, ~ 10^{-12}, and ~ 10^{14}, respectively. Therefore, we can conclude that the influence of Earth's rotation can be completely ignored for the mantle and the inner core, but that fluid motion in the outer core is largely affected by the rotation. Next, the Elsasser number, $\Lambda_E \equiv \sigma B^2/\rho\Omega \approx F/\rho\Omega v$, can be used to evaluate the ratio of the Lorentz force and the Coriolis force. Assuming that $\sigma = 10^6$ Sm^{-1}, the Elsasser number is estimated to be $\Lambda_E = 10^{-1}$–10^3 for $B = 10^{-4}$–10^{-2} T (note that the magnetic flux B is measured by the unit Tesla; the intensity of the magnetic field at Earth's surface is 3×10^{-5} T). The Lorentz force is dominant in the case of a strong magnetic field, while the Coriolis force is important for a weak magnetic field. Viscous resistance is not significant in the fluid motion of the core (except in a very thin boundary layer). In summary, in addition to the pressure gradient and buoyancy, the Coriolis force due to the Earth's rotation as well as the Lorentz force due to the electromagnetic field play important roles in the dynamo process.

Fritz Busse, then at the University of California at Los Angeles, studied the motion of rotating fluid (in the presence of a weak magnetic field, i.e., $\Lambda_E \ll 1$) and showed that the pattern of fluid motion becomes a spiral extending along the rotation axis (fig. 6-5). This type of flow pattern is called the *Taylor column*. The growth of the Taylor column is obstructed by the presence of the inner core. So the Taylor column does not form in the part of in the outer core that corresponds to the northern and southern parts of the inner core. The pattern of fluid motion in the outer core is affected in this way by the inner core. That is, flow pattern is different between the inside and outside of the *tangent cylinder*, which borders the inner core and extends along the rotation axis. Even in quite a large magnetic field, this phenomenon continues to prevail, so the flow pattern represented by the Taylor column is a good starting point to understand the dynamo process (Kageyama and Sato 1997; Olson, Christenson, and Glatzmaier 1999).

The most important concept of the dynamo is the principle of the frozen-in-field theorem found by the Swedish physicist Hannes Alfvén (he

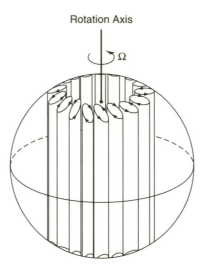

Rotation Axis

Fig. 6-5. Flow in the outer core (after Busse 1975). The flow pattern in the outer core is columnar due to the effect of rotation (the *Taylor column*). This flow is obstructed by the presence of the inner core. Thus, the convection pattern within the core—that is, the dynamo process—is significantly different between the inside and the outside of the *tangent cylinder* (a cylinder that touches the inner-outer core boundary with the axis parallel to the rotation axis). No Taylor column is formed in the tangent cylinder.

was awarded the Nobel Prize for his contribution to magneto-fluid dynamics). According to this principle, when a fluid with high electric conductivity moves, the magnetic field lines also move with the fluid. So, if fluid motion is not uniform, the magnetic field lines will be deformed and the intensity of the magnetic field changes (fig. 6-6). Therefore, a spiral fluid motion like the Taylor column is very efficient in producing the magnetic field. Figure 6-7 shows how a magnetic field will be formed by this kind of fluid motion. Because of the helicity of flow, the poloidal field (a magnetic field with radial and tangential components; fig. 6-8) is transformed to the toroidal field (a magnetic field with no radial components and with closed magnetic field lines within the core), then the toroidal field is in turn transformed into the poloidal field. The magnetic field is maintained by this complicated interaction between the poloidal and toroidal fields. So, even though only the poloidal field can be observed at the surface, there must also be the toroidal field to generate and maintain the (poloidal) magnetic field.

When convection is not so intense, the dynamo involving the fluid flow

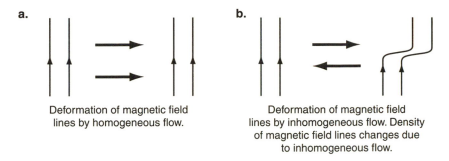

a.

Deformation of magnetic field
lines by homogeneous flow.

b.

Deformation of magnetic field
lines by inhomogeneous flow. Density
of magnetic field lines changes due
to inhomogeneous flow.

Fig. 6-6. Deformation of magnetic field lines by the principle of frozen-in-field magnetic flux.

with the form of the Taylor column operates mainly in the outside of the tangential cylinder. However, vigorous convection is considered to take place in the Earth's outer core, and in this case, an intense dynamo process is expected also within the tangent cylinder. In the model by Peter Olson, at Johns Hopkins University, and others, a large amount of heat is generated in this part, and flow due to the thermal gradient contributes to the formation of a magnetic field (Olson, Christenson, and Glatzmaier 1999).

What kind of magnetic field will be formed from such a model? Due to the geometrical effects of the inner core as described above, the flow in the outer core around the polar region (inside the tangent cylinder) is very different from the flow outside the cylinder. Fluid flow is spiral when the Taylor column is outside the tangent cylinder, whereas it becomes axisymmetric inside the cylinder. In the models by Glatzmaier and Roberts and by Olson and his colleagues, this flow inside the tangent cylinder is intense, and it significantly affects the formation of the magnetic field. Inside the tangent cylinder, upwelling is formed around the poles, and the surrounding fluid flows along with it. Since this flow of the surrounding fluid is not parallel to the rotation vector, it is affected by the Coriolis force. So the flow near the surface of the inner core has the eastward component. On the other hand, flow is westward near the core-mantle boundary. As a result, in the model by Glatzmaier and Roberts, the toroidal field, which is antisymmetric with regard to the equatorial plane, is formed near the poles together with the poloidal dipole field. Any successful dynamo model must reproduce an Earth-like poloidal dipole magnetic field, but the corresponding toroidal field can be very different from model to model. Kuang and Bloxham (1997), for example, obtained a very differ-

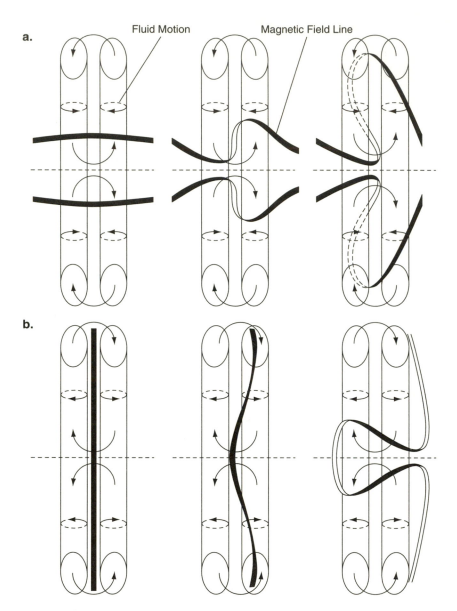

Fluid Motion   Magnetic Field Line

a.

b.

Fig. 6-7. Deformation of magnetic field lines in the outer core due to the flow pattern shown in figure 6-5 (after Olson et al. 1999). Due to the effects of the Coriolis force, the fluid motion has a strong helicity. The magnetic field lines are twisted by the flow with helicity, and (a) the toroidal field is generated from the poloidal field and (b) the poloidal field is generated from the toroidal field. This cycle can reproduce and maintain a magnetic field.

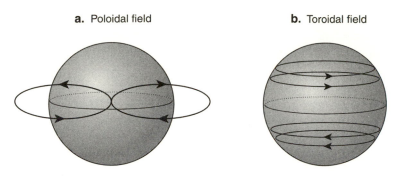

a. Poloidal field   **b.** Toroidal field

Fig. 6-8. The poloidal field (a) and the toroidal field (b).

ent toroidal field by using a slightly different formulation from that used by Glatzmaier and Roberts.

Therefore, one of the keys to understanding the dynamo process is the toroidal magnetic field. In fact, most of differences among the several dynamo models occur in how the poloidal field is reproduced by the toroidal field. While the poloidal field of any model is similar, the size and shape of the toroidal field vary significantly among models. Therefore, if we can infer the toroidal field from some observations, we can judge which model is more realistic. However, the toroidal field, by its definition, does not appear on the surface. So it has been very difficult to investigate the dynamo process based on observations. The situation is changing, however, due to unexpected discoveries in seismology. Some of the seismological observations of the inner core appear to carry important clues to the origin of the geomagnetic field.

6-3. THE DYNAMICS OF THE INNER CORE

6-3-1. Seismological Observations

Earth's core consists of the liquid outer core and the solid inner core. The early study of the dynamo theory used to ignore the existence of the inner core for simplification. Since the 1980s, however, the influence of the inner core has been considered. Two aspects of such influence are described in section 6-2. One is its role as a heat source and the other is its geometrical effect on convection.

In the 1980s and 1990s, seismic observations of the inner core progressed, and revealed that the inner core not only affects convection pattern as a solid sphere, but also contributes to core dynamics in various ways. It has been shown that the inner core has a large seismic anisotropy and also rotates faster than the mantle. As described in chapter 3, seismic anisotropy indicates some anisotropic structure, which implies that the structure of the inner core is somehow dynamically controlled. If the inner core rotates faster than the mantle, some force should be operating on the inner core. These observations are important for core dynamics, but these are delicate observations. Therefore the validity of these observations is still debated, and their implications for core dynamics are controversial.

In 1986, a group of seismologists at Harvard reported unequivocal evidence of elastic anisotropy of the inner core, one result obtained through the analysis of the fine structure of free oscillation (Woodhouse, Giardini, and Li 1986) and the other from the anomalous travel times of body waves that pass through the inner core (Morelli, Dziewonski, and Woodhouse 1986). In these papers, the authors concluded the existence of anisotropy on the basis of the anomalous splitting of spectrum peaks in free oscillations and the azimuthal dependency of body-wave travel time propagating through the inner core. As a first-order approximation, the inner core anisotropy has axial symmetry with regard to the rotation axis. In the direction parallel to the rotation axis, compressional wave velocity is ~ 3% faster than in the direction normal to the rotation axis (fig. 6-9). According to later detailed studies, the symmetry axis is tilted about 10 degrees from the rotation axis. The rotation of the inner core was indeed detected by using this tilted axis of symmetry of anisotropy. Recently it was reported that anisotropy is weak in the upper layer of the inner core and it is significant only at deeper than ~ 200 km. Furthermore, some reported that anisotropy differs greatly between the eastern and western hemispheres (for details see Creager 2000).

In 1996, ten years after the discovery of inner core anisotropy, two groups of seismologists reported that Earth's solid inner core rotates faster than the mantle. This fast rotation of the inner core is sometimes referred to as *super-rotation*. The finding of super-rotation was motivated by the results of geodynamo calculation (Glatzmaier and Roberts 1996). Xiaodong Song and Paul Richards, then at Lamont Doherty Earth Observatory at Columbia University, and Wei-jia Su and his colleagues at Harvard (and the University of California at Berkeley), competed with each other to verify this prediction, and they published similar results almost at the same time in *Nature* and *Science*, respectively (Song and Richards, 1996; Su, Dziewonski, and Jeanloz 1996). Based on the fact that the symmetry

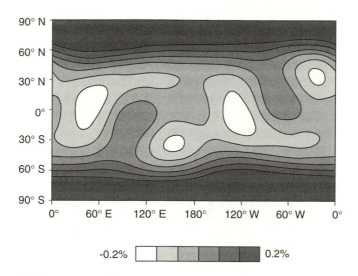

Fig. 6-9. Seismic anisotropy of the inner core based on the study of free oscillations (Woodhouse et al. 1986). Values denote deviations in the frequency of free oscillation peaks. This anisotropy corresponds to about 3% of anisotropy in seismic wave velocity (waves propagating in the direction of the rotation axis are faster by 3%).

axis of anisotropy is tilted from the rotation axis, they investigated the temporal variation of the projected position of the symmetry axis of anisotropy at Earth's surface from seismic records for the past thirty years. They reported that the symmetry axis of anisotropy rotates eastward by 1–3 degrees per year (fig. 6-10). The observations of this inner-core super-rotation have been actively pursued since then, and some different results have also been reported. Some people have cast doubt on the reliability of the initial results. For example, Annie Souriau, at the Observatoire Midi-Pyrénées in Toulous, France, proposed that the uncertainties in the location of earthquakes are too large to allow any meaningful estimation of super-rotation (Souriau 1999). Gabi Laske and Guy Masters, at the Scripps Oceanographic Institution, used the data of anomalies of free oscillations to infer the rate of super-rotation, the results of which are not affected by the uncertainties in the estimation of location of earthquakes (1999). They obtained results of 0.2 ± 0.2 degrees/year. Based on the review of a wide range of observations Ken Creager, at the University of Washington, Seattle, concluded the rate of super-rotation is 0.2 to 0.6 degrees/year (2000).

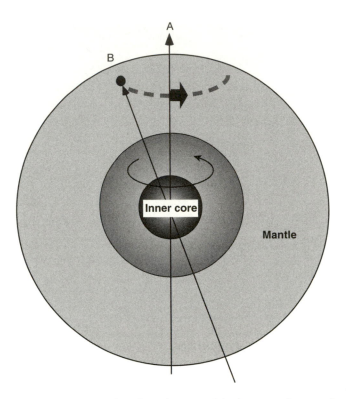

Fig. 6-10. Inner core super-rotation. The anisotropy of the inner core is approximately axisymmetric, but its symmetry axis deviates slightly from the rotation axis. Thus, if the inner core rotates at a rate different from that of the mantle, the symmetry axis of the inner core anisotropy should move with respect to the mantle. This movement was detected by some seismologists by analyzing past seismic records.

6-3-2. The Origin of Anisotropy

Once the inner core anisotropy was discovered, we should then consider its origin. As explained in chapter 3, anisotropy can result from lattice preferred orientation, or layering. For a possible layering origin in the inner core, one may think of the anisotropic melt distribution in partially molten materials. However, according to the analysis by Sumita and his colleagues (1996), melt is likely to be squeezed out from the inner core due to effective compaction. Also, the density and elastic properties of the inner core are consistent with solid iron. So this mechanism is unlikely to be important.

In order to understand the origin of seismic anisotropy due to lattice preferred orientation, two physical properties must be known: the elastic anisotropy of crystals and the nature of the orientation of crystals (lattice preferred orientation) (see chap. 3). Major progress has been made in the study of the elastic properties of iron in the inner core, mainly through quantum mechanical, first-principles calculation. Gerd Steinle-Neumann and his colleagues at the University of Michigan applied an improved technique of quantum mechanical calculations to determine the density and elastic constants of iron in the inner core (Steinle-Neumann et al. 2001). Iron is a well-known material and usually assumes the body-centered-cubic (bcc) structure (called α-iron) at room pressure and room temperature. If iron is heated to more than ~ 1,200 K (at room pressure), then its crystal structure changes to the face-centered-cubic (fcc) structure (γ-iron). However, if iron is under high pressure, it changes to another structure, the hexagonal-close-pack (hcp) structure (ε-iron). This hcp iron (ε-iron) is believed to be the main constituent of Earth's inner core. All materials with the hcp structure are highly anisotropic (Zn, Ti, etc.). Consequently, iron in the inner core is expected to be anisotropic. However, it took some time to understand the nature of the anisotropy of ε-iron in Earth's inner core. Direct determination of elastic anisotropy under inner-core conditions is still not possible. Therefore, the only plausible way to infer the elastic properties of the inner core is through quantum mechanical calculations. Lars Stixrude, then at the Carnegie Institution of Washington, D.C., and Ron Cohen made the first calculation of this kind ignoring, to a good approximation as many thought, the effects of temperature (Stixrude and Cohen 1995). They found rather weak anisotropy and therefore proposed that the inner core might be made of a single crystal of iron. However, when Steinle-Neumann and his colleagues (2001) later extended their work to higher temperatures, they found that the nature of anisotropy changes dramatically with temperature. Not only does the magnitude of anisotropy increase with temperature, but the geometry of anisotropy is also completely different from that at zero K: along the c-axis the seismic wave velocity is the slowest at high pressures and high temperatures, whereas the seismic wave velocity is the fastest along the c-axis at high pressures and low temperatures. Such a strong effect of temperature is anomalous, but in this case it is related to the change in crystal structure: the spacing of closed-pack planes (i.e., the length of the c-axis) gets larger as temperature increases, and consequently the stiffness along the c-axis becomes weaker.

Given the elastic anisotropy of individual crystals, we then need to know how they could be aligned. This part of the model for seismic anisotropy in the inner core is still highly controversial and there is yet no widely accepted theory. Macroscopic processes for anisotropic structure formation in the inner core are likely different from those for mantle anisotropy because the physical environment in the inner core is different from that in the mantle. First, the inner core is the product of growth (through solidification) from the molten outer core. Second, because of the high thermal conductivity of core materials and because of the low concentration of radiogenic elements, it is unlikely that thermal convection occurs in the inner core. Third, the strength of the magnetic field is much higher in the inner core than at the surface of Earth. Therefore, processes other than familiar thermal convection might be responsible for anisotropic structure formation. Any models of anisotropy must be consistent with the two observations: anisotropic structure has approximate axial symmetry around the rotation axis, and the anisotropy is weak in the shallow regions.

There are two groups of models for lattice preferred orientation. One is by deformation, and the other is by grain growth (recrystallization). In the case of deformational preferred orientation, the strain must be large enough (about the order of one or more). Furthermore, the microscopic mechanism of deformation cannot be diffusion creep; it has to be dislocation creep or twinning (chap. 3). In the case of grain growth, some anisotropic field must exist sufficiently long for seismic anisotropy to grow.

Soon after the anisotropy of the inner core was published, Raymond Jeanloz and Rudy Wenk, at the University of California at Berkeley, proposed a model to explain it (1988). They argued that thermal convection occurs in the inner core and deformation-induced texturing (lattice preferred orientation) causes seismic anisotropy. The proposed macroscopic mechanism for anisotropic structure formation, thermal convection, is much the same as the way anisotropy is created in the mantle (chap. 3). However, as described previously, thermal convection hardly occurs in the inner core. Moreover, this model cannot explain the axial symmetry of anisotropy (because of the very small Taylor number for the inner core, the rotation does not have any effects for the solid state flow in the inner core).

By taking these drawbacks into account, another model was proposed, in which anisotropy is explained by grain growth, not by ductile deformation. When the inner core grows by the crystallization of solid iron, lattice preferred orientation may develop due to the existence of an anisotropic field. I considered the anisotropy of the geomagnetic field (Karato 1993a), and Michael Bergman considered the anisotropy of heat

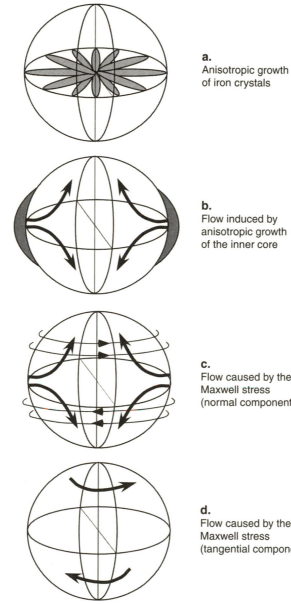

a.
Anisotropic growth
of iron crystals

b.
Flow induced by
anisotropic growth
of the inner core

c.
Flow caused by the
Maxwell stress
(normal component)

d.
Flow caused by the
Maxwell stress
(tangential component)

Fig. 6-11. Mechanisms of seismic anisotropy in the inner core: (a) Anisotropic growth of Fe crystals due to anisotropy in heat transfer (Bergman 1997) or a magnetic field (Karato 1993); (b) Anisotropy caused by stress-induced lattice preferred orientation due to flow associated with anisotropic growth of the inner core (Yoshida et al. 1996); (c) Anisotropy due to lattice preferred orientation caused by flow due to normal components of the magnetic field (Karato 1999); (d) Anisotropy due to lattice preferred orientation caused by flow due to the tangential component of the magnetic field (Buffett and Wenk 2001).

flux (Bergman 1997) (fig. 6-11a). Because these fields have rotational symmetry, the rotational symmetry of seismic anisotropy can be explained naturally. However, these models have two common problems. First, if anisotropy is acquired at the surface, then anisotropy must be stronger near the surface than in the deeper portions. This is not consistent with the seismological observations that suggest that the shallow layer is nearly isotropic. Second, even though anisotropic structures may be formed by growth, subsequent deformation would modify it if deformation with large strain occurs in the inner core.

Shigeo Yoshida and his colleagues, then at the University of Tokyo, proposed a noble idea for the origin of the inner-core anisotropy (Yoshida, Sumita, and Kumazawa 1996) (fig. 6-11b). They considered the anisotropic growth of the inner core as a macroscopic mechanism to create the anisotropic structure. The inner core grows by the cooling and resulting solidification of the outer core. The rate of growth is controlled by the heat flow in the outer core, but the efficiency of heat transfer in the outer core differs between the polar and the equatorial directions. Because of the Taylor column (fig. 6-5) presumably present in the outer core, heat transfer is more efficient in the equatorial region. Thus, the growth of the inner core is faster in the equatorial region than in the polar regions. As a result, the inner core should have a swell around the equator. The gravity force acts on a swell around the equator, which results in flow. In a steady state, the velocity of this flow corresponds to the rate of the inner-core growth. This flow induces a deviatoric stress field within the inner core. Because of the stress field generated by this flow, iron crystals in the inner core have different energy for different orientations. The selective growth of crystals with low-energy orientation would then lead to lattice preferred orientation. Although this model provides an elegant explanation for the axisymmetric anisotropy, it has a serious problem: the driving force for this process is so small that it is hard to produce a strong enough lattice preferred orientation within the age of Earth for a reasonable range of viscosity.

Bruce Buffett in Canada proposed that the gravitational interaction between the mantle and core might cause deformation leading to seismic anisotropy (2000). As can be seen from the results of seismic tomography, density heterogeneity is likely to be present in the mantle. The solid inner core must deform by the force due to this density heterogeneity. When the inner core rotates with respect to the mantle, as will be discussed later, the inner core should always be displaced from its equilibrium shape. Thus, flow has to take place to bring the inner core to dynamic equilibrium. He asserted that this deformation and the accompanying stress might cause anisotropy. Although the advantage of this mechanism is that it can create quite a large force ($\sim 10^3$ Pa), the problem is that the strain is very small ($\sim 10^{-4}$). This is because this deformation results as relaxation toward equilibrium, so when the system reaches equilibrium, no more strain can be accumulated. Consequently, this mechanism can hardly produce an anisotropic structure.

How about the other forces—that is, the electromagnetic force? Inside the core, the strength of the magnetic field is higher than that at the surface. Can this magnetic field contribute to the flow in the inner core, helping to generate an anisotropy structure? The intensity of the electromagnetic force depends on the strength of a magnetic field, and the electromagnetic stress is on the order of B^2/μ_0 (B is a magnetic-flux density, and μ_0 is the magnetic permeability of the vacuum). Assuming that the strength of the magnetic field is $\sim 10^{-2}$ T (by extrapolating the surface values, the strength of the poloidal field in the inner core is $\sim 10^{-3}$ T. Most models of the geodynamo show that the toroidal field is stronger than the poloidal field), the magnitude of electromagnetic stress (the Maxwell stress) is estimated to be $\sim 10^2$ Pa. If the viscosity of the inner core is $\sim 10^{17}$ Pa·s, the corresponding strain rate is calculated to be $\sim 10^{-15}$/s. Although this stress is somewhat smaller than the maximum magnitude of stress due to gravitational coupling between the mantle and core, it may generate a larger strain because it can last as long as a magnetic field is present. Obviously, the symmetry of the strain field is controlled by the symmetry of the magnetic field that is approximately symmetric with respect to the rotational axis. Deformation by this force is, therefore, a plausible cause for the anisotropic structure.

Two different models have been proposed. Karato (1999) emphasized the role of normal stress caused by the magnetic field. The normal stress is developed by the magnetic field that varies with the latitude. Consequently, a radial flow pattern will develop inside the inner core (fig. 6-11c). A different version of magnetic-field-induced flow model was proposed by Buffett and Wenk (2001) (fig. 6-11d) who emphasized the role of the tangential component of stress. Both models roughly reproduce the observed seismic anisotropy if the latest computational results of elastic anisotropy are used, although the depth variation of the flow is different. The model by Buffett and Wenk (2001) results in a strong flow near the surface but a weak flow in the deep portions, whereas the model by Karato (1999) results in the strong field in the deep portions. One of the interesting features of these magnetic-field-related models is that the nature of seismic anisotropy is closely related to the geometry (and strength) of the magnetic field, particularly the toroidal field. As will be shown below, the possible cause for super-rotation is also the magnetic coupling between the inner and the outer core, and in this case, the two observations will have the same origin. Therefore, in these models, there is a link between the magnetic field and the seismological observations, and in principle, seismological observations could provide constraints on the magnetic field.

6-3-3. The Origin of Inner-Core Super-Rotation

The inner-core super-rotation should originate in torque working on the inner core. Glatzmaier and Roberts (1995a, 1996), who first predicted the inner-core super-rotation, proposed that the magnetic interaction between the outer and the inner core causes the super-rotation. According to their model, the magnetic coupling exerts a torque to the inner core, and this torque will make a balance with the torque that is due to viscous coupling. Because the magnitude of the magnetic torque depends on the magnitude of toroidal field, the magnitude of toroidal field can be inferred from the magnitude of super rotation. Jon Aurnou and his colleagues, then at Johns Hopkins University (Aurnou, Brito, and Olson 1996) estimated the strength of the toroidal field in this manner. In fact, the magnitude and geometry of toroidal field is highly dependent on the model. Kuang and Bloxham's model (1997) shows a quite different toroidal field from that of Glatzmaier and Roberts, and Kuang and Bloxham predicted a much smaller rate of inner-core rotation. Therefore, the inferred rate of super-rotation provides a constraint on the models of the geodynamo.

6-4. SUMMARY AND OUTLOOK

The core accounts for more than one-third of the total mass of Earth. However, because the core pressures and temperatures are difficult to reproduce in laboratories, and because the study of the core has little influence on other fields except geomagnetism, the dynamics of the core was a research topic of interest only to experts in geomagnetism.

In recent few years, the study of Earth's core has made remarkable progress. This area is becoming highly interdisciplinary. It is remarkable that the realistic geomagnetic field has been reproduced by fully dynamic computer simulations. The driving force for this progress is numerical calculations with powerful computers as well as the seismological study of the core structure, particularly of the inner-core structure. The study of the core materials with theoretical and experimental methods also significantly contributed to this progress. Seismological studies, especially, have changed the scope of the study of the core. The observations of the geomagnetic field used to be the only observational constraint on the core. Now we have seismological observations as well.

Most of seismological observations regard the solid inner core. Solids usually exhibit more structure than liquids (i.e., the outer core), especially because solids often have anisotropy. Interpretations of these observations are still controversial, particularly in the case of establishing the cause for anisotropy. To develop a new theory for the origin of geomagnetism on the basis of seismic observations, we need the better understanding of macroscopic processes that may be operating in the core as well as microscopic properties of core materials, such as the deformation mechanisms.

One of the interesting aspects of core dynamics is that strong interaction has been revealed among various layers such as the outer core, the inner core, and the mantle. Though this chapter deals mainly with interaction between the outer core and the inner core, the interaction between the core and the mantle is also important. The cooling rate of the core is determined by heat transfer through the mantle, and it is likely that the core structure (e.g., the existence of the inner core) is determined by this interaction.

Although there have been major breakthroughs in the study of Earth's core, there remain a number of fundamental questions. First, although the initial successful geodynamo calculations have been made, the validity of these calculations is still debated. One of the limitations of these models is the treatment of the mechanical boundary conditions in the outer core. Because the fluid viscosity of the outer core is very low, there will be a very thin mechanical boundary layer between the outer core and the mantle and the inner core. This mechanical boundary layer is referred to the *Ekman boundary layer*, whose thickness is estimated to be ~ 1 m or less. There is no easy way to include this boundary layer in numerical calculations. Various tricks have been used in numerical modeling to overcome this difficulty, including the use of fictitious high viscosity of the outer-core materials. Second, although the magnetic field reversals have also been documented by numerical simulations (Glatzmaier and Roberts 1995b), the details of reversals remain unknown. Glatzmaier and his colleagues (1999) noticed that when a realistic thermal boundary condition is imposed at the core-mantle boundary, unrealistic reversal behavior arises: the magnetic field becomes too unstable. They suggested that the temperature variations at the core-mantle boundary might be much more homogeneous than a simple model predicts and proposed that the D'' layer is nearly isothermal (little lateral variation in temperature). However, the validity of such a notion is not clear from the geodynamic point of view. The effects of boundary conditions at the inner-outer core boundary may need to be investigated in more detail. Third, the seismological measurements of both anisotropy and super-rotation are difficult to obtain and the results are still debated. In particular the validity of initial observations of super-rotation has been questioned. The current observations point to much lower values; $\sim 0.2-0.6$ degree/year or less (Creager 2000). If indeed there is no super-rotation, then one would have a constraint on the inner core viscosity such that the viscosity is between $\sim 10^{16}$ and $\sim 10^{20}$ Pa·s (Buffett 1997). Fourth, the observed sharp transition from isotropic (near surface) layer to strongly anisotropic layer needs a physical explanation. This transition appears to occur at different depths between the eastern and western hemispheres. The presence of these difficulties or challenges clearly indicates that the dynamics of this central portion of Earth still remain enigmatic, although in a much more refined way than a few years ago.

REFERENCES

Agee, C. B. 1993. Petrology of the mantle transition zone. *Ann. Rev. Earth Planet. Sci.* 21:19–42.

Aki, K. 1968. Seismological evidence for the existence of soft thin layers in the upper mantle beneath Japan. *J. Geophys. Res.* 73:585–94.

Aki, K., Christoffersson, A., and Husebye, F. S. 1977. Determination of three-dimensional seismic structure of the lithosphere. *J. Geophys. Res.* 82:277–96.

Allègre, C. J. 1997. Limitation on the mass exchange between the upper and lower mantle: The evolving convection regime of the Earth. *Earth Planet. Sci. Lett.* 150:1–6.

Allègre, C. J., and Turcotte, D. L. 1986. Implications of a two-component marble-cake mantle. *Nature* 323:123–27.

Ando M., Ishikawa, Y., and Wada, H. 1980. S-wave anisotropy in the upper mantle under a volcanic area in Japan. *Nature* 268:43–46.

Anderson, D. L. 1989. *Theory of the Earth*. Boston: Blackwell.

Anderson, D. L., Tanimoto, T., and Zhang, Y-S. 1992. Plate tectonics and hot spots—The third dimension. *Science* 256:1645–51.

Anderson, O. L. 2002. The three-dimensional phase diagram of iron. In *Core Dynamics: Structure and Rotation,* edited by V. Dehant, K. C. Creager, S. Zatman, and S. Karato. Washington, D.C.: Amer. Geophys. Union, In press.

Aurnou, J. M., Brito, D., and Olson, P. L. 1996. Mechanics of inner core super-rotation. *Geophys. Res. Lett.* 23:3401–04.

Bai, Q., and Kohlstedt, D. L. 1993. Effects of chemical environment on the solubility and incorporation mechanism for hydrogen in olivine. *Phys. Chem. Mineral.* 19:460–71.

Becker, T. W., Kellogg, J. B., and O'Connell, R. J. 1999. Thermal constraints on the survival of primitive blobs in the lower mantle. *Earth Planet. Sci. Lett.* 171:351–65.

Benioff, H. 1949. Seismic evidence for the fault origin of ocean deeps. *Geol. Soc. Amer. Bull.* 60:1837–56.

Benz, W., Slattery, W. L., and Cameron, A. G. W. 1989. The origin of the Moon and the single-impact hypothesis, pt. II. *Icarus* 71:30–45.

Bercovici, D., Schubert, G., and Tackley, P. J. 1993. On the penetration of the 660 km phase change by mantle downflows. *Geophys. Res. Lett.* 20:2599–2602.

Bergman, M. I. 1997. Measurements of elastic anisotropy due to solidification texturing and the implications for the Earth's inner core. *Nature* 389:60–63.

Bijwaard, H., and Spakman, W. 1999. Tomographic evidence for a narrow whole mantle plume below Iceland. *Earth Planet. Sci. Lett.* 166:121–26.

Bijwaard, H., Spakman, W., and Engdahl, E. R. 1998. Closing the gap between regional and global tomography. *J. Geophys. Res.* 103:30055–78.

Birch, F. 1952. Elasticity and constitution of the Earth's interior. *J. Geophys. Res.* 57:227–86.

———. 1961. The velocity of compressional waves in rocks to 10 kilobars, pt. 2. *J. Geophys. Res.* 66:2199–2224.

Blacic, J. D. 1972. Effect of water on the experimental deformation of olivine. In *Flow and Fracture of Rocks*, edited by H. C. Heard, I. Y. Borg, N. L. Carter, and C. B. Raleigh. pp. 109–15. Washington D.C.: Amer. Geophys. Union.

Bolton, H. 1996. Long period travel times and the structure of the mantle. Ph.D. diss. University of California at San Diego.

Braginski, S. I. 1963. Structure of the F layer and reasons for convection in the Earth's core. *Dokl. Akad. Nauk SSSR Engl. Transl.* 149:1311–14.

Breuer, D., and Spohn, T. 1995. Possible flush instability in mantle convection at the Archean-Proterozoic transition. *Nature* 378:608–10.

Bridgman, P. W. 1945. Polymorphic transitions and geological phenomena. *Amer. J. Sci.* 243A:90–97.

Buffett, B. A. 1997. Geodynamic estimate of the viscosity of the Earth's inner core. *Nature* 388:571–73.

———. 2000. Dynamics of the Earth's core. In *Earth's Deep Interior*, edited by S. Karato, A. M. Forte, R. C. Liebermann, G. Maters, and L. Stixrude, pp. 37–62. Washington, D.C.: Amer. Geophys. Union.

Buffett, B. A., Garnero, E. J., and Jeanloz, R. 2000. Sediments at the top of Earth's core. *Science* 290:1338–42.

Buffett, B. A., Huppert, H. E., Lister, J. R., and Woods, A. W. 1992. Analytical model for solidification of the Earth's core. *Nature* 356:329–31.

Buffett, B. A., and Wenk, H-R. 2001. Texturing of the Earth's inner core by Maxwell stress. *Nature* 413:60–63.

Bullard, E. C. 1949. Magnetic fields within the earth. *Proc. Roy. Soc. Lond.* A197:433–53.

Busse, F. H. 1975. Patterns of convection in spherical shells. *J. Fluid. Mech.* 72:67–85.

Buttles, J., and Olson, P. 1998. A laboratory model of subduction zone anisotropy. *Earth Planet. Sci. Lett.* 164:245–62.

Cathles, L. M. 1975. *The Viscosity of the Earth's Mantle*. Princeton, N.J.: Princeton University Press.

Chopra, P. N., and Paterson, M. S. 1984. The role of water in the deformation of dunite. *J. Geophys. Res.* 89:7861–76.

Christensen, U. R., and Hofmann, A. W. 1994. Segregation of subducted oceanic crust in the convecting mantle. *J. Geophys. Res.* 99:19867–884.

Christensen, U. R., and Yuen, D. A. 1984. The interaction of a subducting lithospheric slab with a chemical or phase boundary. *J. Geophys. Res.* 89:4389–4402.

————. 1985. Layered convection induced by phase transitions. *J. Geophys. Res.* 90:10291–300.

Connerney, J. E. P., Acuña, M. H., Wasilewski, P. J., Ness, F. P., Rème, H., Mazelle, C., Vignes, D., Lin, R. P., Mitchell, D. L., and Cloutier, P. A. 1999. Magnetic lineations in the ancient crust of Mars. *Science* 284:794–98.

Creager, K. C. 2000. Inner core anisotropy and rotation. In *Earth's Deep Interior*, edited by S. Karato, A. M. Forte, R. C. Liebermann, G. Maters, and L. Stixrude, pp. 89–114. Washington, D.C.: Amer. Geophys. Union.

Creager, K. C., and Jordan, T. H., 1974. Slab penetration into the lower mantle. *J. Geophys. Res.* 79:3031–49.

Davies, G. F. 1995. Penetration of plates and plumes through the mantle transition zone. *Earth Planet. Sci. Lett.* 133:507–16.

de Smet, J. H., van den Berg, A. P., and Vlaar, N. J. 1998. Stability and growth of continental shields in mantle convection models including recurrent melt production. *Tectonophysics* 296:15–29.

Doin, M-P., Fleitout, L., and Christensen, U. R. 1997. Mantle convection and stability of depleted and undepleted continental lithosphere. *J. Geophys. Res.* 102:2771–87.

Dupas-Bruzek, C., Sharp, T. G., Rubie, D. C., and Durham, W. B. 1998. Mechanisms of transformation and deformation in $Mg_{1.8}Fe_{0.2}SiO_4$ olivine and wadsleyite under non-hydrostatic stress. *Phys. Earth Planet. Inter.* 108:33–48.

Dziewonski, A. M. 1984. Mapping the lower mantle: Determination of lateral heterogeneity in P velocity up to degree and order 6. *J. Geophys. Res.* 89:5929–52.

Dziewonski, A. M., and Anderson, D. L. 1981. Preliminary reference Earth model. *Phys. Earth Planet. Inter.* 25:297–356.

Dziewonski, A. M., Hager, B. H., and O'Connell, R. J. 1977. Large-scale heterogeneities in the lower mantle. *J. Geophys. Res.* 82:239–55.

Ekström, G., and Dziewonski, A. M. 1998. The unique anisotropy of the Pacific upper mantle. *Nature* 394:168–72.

Elsasser, W. M. 1946. Induction effects in terrestrial magnetism, pt. 1: Theory. *Phys. Rev.* 70:202–12.

Fearn, D. R., Loper, D. E., and Roberts, P. H. 1981. Structure of the earth's inner core. *Nature* 292:232–33.

Fischer, K. M., Fouch, M. J., Wiens, D. G., and Boettcher, M. S. 1998. Anisotropy and flow in Pacific subduction zone back-arcs. *PAGEOPH* 151:463–75.

Foulger, G. R., Prithard, M. J., Julian, B. R., Evans, J. R., Allen, R. M., Nolet, G., Morgan, W. J., Bergasson, B. H., Erlendsson, P., Jakpbsdottier, S., Ragnarsson, S., Stefansson, R., and Vogfjörd, K. 2001. Seismic tomography shows that upwelling beneath Iceland is confined to the upper mantle. *Geophys. J. Int.* 146:504–30.

Forsyth, D. W. 1975. The early structural evolution and anisotropy of the oceanic upper mantle. *Geophys. J. Roy. Astr. Soc.* 43:103–62.

Forsyth, D. W., and Uyeda, S. 1975. On the relative importance of driving forces of plate motion. *Geophys. J. R. Astr. Soc.* 43:163–200.

Frolich, C. 1989. The nature of deep earthquakes. *Ann. Rev. Earth Planet. Sci.* 54:227–54.

Fukai, Y. 1984. The iron-water interaction and the evolution of the Earth. *Nature* 308:174–75.

Fukao, Y., Obayashi, M., Inoue, H., and Nembai, M. 1992. Subducting slabs stagnant in the mantle transition zone. *J. Geophys. Res.,* 97:4809–22.

Fukao, Y., Widiyantoro, S., and Obayashi, M. 2001. Stagnant slabs in the upper and lower mantle transition zone. *Rev. Geophys.* 39:291–323.

Gaherty, J. B. 2001. Seismic evidence for hotspot-induced buoyant flow beneath the Reykjanes Ridge. *Science* 293:1645–47.

Gaherty, J. B., and Hager, B. H. 1994. Compositional versus thermal buoyancy and the evolution of subducted lithosphere. *Geophys. Res. Lett.* 21:141–44.

Gaherty, J. B., and Jordan, T. H. 1995. Lehmann discontinuity as the base of an anisotropic layer beneath continents. *Science* 268:1468–71.

Garnero, E. J., Revenaugh, J., Williams, Q., Lay, T., and Kellogg, L. H. 1998. Ultralow velocity zone at the core-mantle boundary. In *The Core-Mantle Boundary Regions,* edited by M. Gurnis, M. E. Wysession, E. Knittle, and B. A. Buffett, pp. 319–34. Washington, D.C.: Amer. Geophys. Union.

Glatzmaier, G. A., and Roberts, P. H. 1995a. A three-dimensional convective dynamo solution with rotating and finitely conducting inner core and mantle. *Phys. Earth Planet. Inter.* 91:63–75.

———. 1995b. A three-dimensional self-consistent computer simulation of a geomagnetic reversal. *Nature* 377:203–8.

———. 1996. Rotation and magnetism of Earth's inner core. *Science* 274:1887–91.

Glatzmaier, G. A., Coe, R. S., Hongre, L., and Roberts, P. H. 1999. The role of the Earth's mantle in controlling the frequency of geomagnetic reversals. *Nature* 401:885–90.

Goetze, C., and Evans, B. 1979. Stress and temperature in the bending lithosphere as constrained by experimental rock mechanics. *Geophys. J. R. Astr. Soc.* 59:463–78.

Grand, S. P. 1994. Mantle shear structure beneath the Americas and surrounding oceans. *J. Geophys. Res.* 99:11591–621.

Green, H. W., II. 1994. Solving the paradox of deep earthquakes. *Sci. Amer.* 271:64–71.

Green, H. W., II, and Burnley, P. C. 1989. A new self-organizing mechanism for deep earthquakes. *Nature* 341:733–37.

Green, H. W., II, and Houston, H. 1995. The mechanics of deep earthquakes. *Ann. Rev. Earth Planet. Sci.* 23:169–213.

Green, H. W., II, Young, T. E., Walker, D., and Scholtz, C. 1990. Anticrack-associated faulting at very high pressure in natural olivine. *Nature* 348:720–22.

Griggs, D. T., and Baker, D. W. 1969. The origin of deep focus earthquakes. In

Properties of Matter under Unusual Conditions, edited by H. Mark and S. Fern-
bach. pp. 23–42. New York: Wiley.

Griggs, D. T., and Blacic, J. D. 1965. Quartz: Anomalous weakness of synthetic
crystals. *Science* 147:292–95.

Griggs, D. T., and Handin, J. 1960. Observations on fracture and a hypothesis of
earthquakes, In *Rock Deformation,* edited by D. T. Griggs and J. Handlin,
pp. 347–73. Geol. Soc. Amer. Memoir 79. Washington, D.C.: Geological Society of America.

Grossman, L. 1972. Condensation in the primitive solar nebula. *Geochim. Cosmochim. Acta* 36:597–619.

Gueguen, Y., and Mercier, J. M. 1973. High attenuation and low-velocity zone in
the upper mantle. *Phys. Earth Planet. Inter.* 7:39–46.

Guillou-Frottier, L., Buttles, J., and Olson, P. 1996. Laboratory experiments on
the structure of subducted lithosphere. *Earth Planet. Sci. Lett.* 133:19–34.

Gurnis, M., Wysession, M. E., Knittle, E., and Buffett, B. A. 1998. *The Core-
Mantle Boundary Region.* Washington D.C.: Amer. Geophys. Union.

Gutenberg, B. 1954. Low-velocity layers in the Earth's mantle. *Geol. Soc. Amer.
Bull.* 65:337–47.

Hager, B. H. 1984. Subducted slabs and the geoid—Constraints on mantle rheology and flow. *J. Geophys. Res.* 89:6003–15.

Hager, B. H., Clayton, R. W., Richards, M. A., Comer, R. P., and Dziewonski, A.
M. 1985. Lower mantle heterogeneity, dynamic topography, and the geoid. *Nature* 313:541–45.

Hart, S. R. 1988. Heterogeneous mantle domains: Signatures, genesis, and mixing chronologies. *Earth Planet. Sci. Lett.* 90:273–96.

Haskell, N. A. 1935. The motion of a viscous fluid under a surface load. *Physics*
6:265–69.

Hedlin, M. A. H., Shearer, P. M., and Earle, P. S. 1997. Seismic evidence of small-
scale heterogeneity throughout the Earth's mantle. *Nature* 387:145–50.

Hess, H. 1964. Seismic anisotropy of the uppermost mantle beneath oceans. *Nature* 203:629–31.

Hirth, G., Evans, R. L., and Chave, A. D. 2000. Comparison of continental and
oceanic mantle electrical conductivity: Is the Archean lithosphere dry? *Geochem.
Geophys. Geosys.,* paper no. 2000GC000048.

Hirth, G., and Kohlstedt, D. L. 1996. Water in the oceanic upper mantle: Implications for rheology, melt extraction, and the evolution of the lithosphere. *Earth
Planet. Sci. Lett.* 144:93–108.

Hobbs, B. E., and Ord, A. 1988. Plastic instabilities: Implications for the origin
of intermediate and deep focus earthquakes. *J. Geophys. Res.* 93:10521–40.

Hofmann, A. W. 1997. Mantle geochemistry: The message from oceanic volcanism. *Nature* 385:219–29.

Honda, H. 1932. On the type of the seismograms and the mechanism of deep
earthquakes. *Geophys. Mag.* 5:301–24.

Honda, S. 1995. A simple parameterized model of Earth's thermal history with the transition from layered to whole mantle convection. *Earth Planet. Sci. Lett.* 131:357–70.

Honda, S., Yuen, D. A., Balachandar, S., and Reuteler, D. 1993. Three-dimensional mantle dynamics with an endothermic phase transition. *Science* 259:1308–11.

Iidaka, T., and Suetsugu, D. 1992. Seismological evidence for metastable olivine inside a subducting slab. *Nature* 356:593–95.

Irifune, T., Kuroda, K., Funamori, N., Uchida, T., Yagi, T., Inoue, T., and Miyajima, N. 1996. Amorphization of serpentine at high pressure and high temperature. *Science* 272:1468–70.

Irifune, T., and Ringwood, A. E. 1987. Phase transformations in primitive MORB and pyrolite composition to 25 GPa and some geophysical implications. In *High-Pressure Research in Mineral Physics,* edited by M. H. Manghnani and Y. Syono, pp. 231–42. Washington, D.C.: Amer. Geophys. Union.

Isaak, D. G., Anderson, O. L., and Cohen, R. E. 1992. The relationship between shear and compressional velocities at high pressures: Reconciliation of seismic tomography and mineral physics. *Geophys. Res. Lett.* 19:741–44.

Isacks, B., and Molnar, P. 1971. Distribution of stresses in the descending lithosphere from a global survey of focal-mechanism solutions of mantle earthquakes. *Rev. Geophys. Space Phys.* 9:103–74.

Ishii, M., and Tromp, J. 1999. Normal-mode and free-air gravity constraints on lateral variations in velocity and density of Earth's mantle. *Science* 285:1231–36.

Ito, E., and Katsura, T. 1989. A temperature profile of the upper mantle transition zone. *Geophys. Res. Lett.* 16:425–528.

Ito, E., and Sato, H. 1990. Aseismicity in the lower mantle by superplasticity in a subducting slab. *Nature* 351:140–41.

Ito, E., and Yamada, H. 1982. Stability relations of silicate spinels, ilmenites, and perovskites. In *High Pressure Research in Geophysics,* edited by S. Akimoto and M. H. Manghnani, pp. 405–19. Tokyo: Terra Pub.

Jackson, I., Paterson, M. S., and FitzGerald, J. D. 1992. Seismic wave dispersion and attenuation in Åheim dunite: An experimental study. *Geophys. J. Int.* 108:517–34.

Jeanloz, R., and Knittle, E. 1989. Density and composition of the lower mantle. *Phil. Trans. Roy. Soc. London* A328:377–89.

Jeanloz, R., and Wenk, H-R. 1988. Convection and anisotropy of the inner core. *Geophys. Res. Lett.* 15:72–75.

Jones, J. H., and Drake, M. J. 1986. Geochemical constraints on core formation in the Earth. *Nature* 322:221–28.

Jordan, T. H. 1975. The continental tectosphere. *Rev. Geophys. Space Phys.* 13:1–12.

———. 1977. Lithospheric slab penetration into the lower mantle beneath the Sea of Okhotsk. *J. Geophys. Res.* 82:473–96.

Jung, H., and Karato, S. 2001. Water-induced fabric transitions in olivine. *Science,* 293:1460–63.

Justice, M. G., Jr., Graham, E. K., Tressler, R. E., and Tsong, I. S. T. 1982. The effect of water on high-temperature deformation in olivine. *Geophys. Res. Lett.* 9:1005–8.

Kageyama, A., and Sato, T. 1997. Generation mechanisms of a dipole field by a magnetohydrodynamical dynamo. *Phys. Rev. E.* 55:4617–26.

Kanamori, H., and Anderson, D. L. 1977. Importance of physical dispersion in surface wave and free oscillation problems. *Rev. Geophys. Space Phys.* 15:105–12.

Kanamori, H., Anderson, D. L., and Heaton, T. H. 1998. Frictional melting during the rupture of the 1994 Bolivian earthquake. *Science* 279:839–42.

Karato, S. 1986. Does partial melting reduce the creep strength of the upper mantle? *Nature* 319:309–10.

———. 1990. The role of hydrogen in the electrical conductivity of the upper mantle. *Nature* 347:272–73.

———. 1993a. Anisotropy of Earth's inner core due to magnetic-field-induced preferred orientation of iron. *Science* 262:1708–11.

———. 1993b. Importance of anelasticity in the interpretation of seismic tomography. *Geophys. Res. Lett.* 20:1623–26.

———. 1995. Effects of water on the seismic wave velocities of upper mantle. *Proc. Japan Acad.,* ser. B, 71:61–66.

———. 1997. On the separation of crustal component from subducted oceanic lithosphere near the 660-km discontinuity. *Phys. Earth Planet. Inter.* 99:103–11.

———. 1998. Seismic anisotropy in the deep mantle, boundary layers, and the geometry of mantle convection. *PAGEOPH* 151:565–87.

———. 1999. Seismic anisotropy in Earth's inner core resulting from Maxwell stresses. *Nature* 402:871–73.

———. 2002. Mapping water content in the upper mantle. In *The Subduction Factory,* edited by J. M. Eiler and G. Abers. Washington, D.C.: Amer. Geophys. Union. In press.

Karato, S., and Jung, H. 1998. Water, partial melting, and the origin of seismic low velocity and high attenuation in the upper mantle. *Earth Planet. Sci. Lett.* 157:193–207.

———. 2002. Effects of pressure on dislocation creep in olivine. *Philos. Mag.* In press.

Karato, S., and Karki, B. B. 2001. Origin of lateral variation of seismic wave velocities and density in the deep mantle. *J. Geophys. Res.* 106:21771–83.

Karato, S., Dupas-Bruzek, C., and Rubie, D. C. 1998. Plastic deformation of silicate spinel under the transition zone conditions. *Nature* 395:266–69.

Karato, S., and Murthy, V. R. 1997. Core formation and chemical equilibrium in the Earth, pt. 1: Physical considerations. *Phys. Earth Planet. Inter.* 100:61–79.

Karato, S., Paterson, M. S., and FitzGerald, J. D. 1986. Rheology of synthetic olivine aggregates: Influence of grain size and water. *J. Geophys. Res.* 91:8151–76.

Karato, S., Riedel, M. R., and Yuen, D. A. 2001. Rheological structure and deformation of subducted slabs in the mantle transition zone: Implications for mantle circulation and deep earthquakes. *Phys. Earth Planet. Inter.* 127:83–108.

Karato, S., and Rubie, D. C. 1997. Toward an experimental study of deep mantle rheology—A new multianvil sample assembly for deformation experiments under high pressures and temperatures. *J. Geophys. Res.* 102:20111–22.

Karato, S., Wang, Z., Liu, B., and Fujino, K. 1995. Plastic deformation of garnets: Systematics and implications for the rheology of the mantle transition zone. *Earth Planet. Sci. Lett.* 130:13–30.

Karato, S., Zhang, S., and Wenk, H-R. 1995. Superplasticity in the Earth's lower mantle: Evidence from seismic anisotropy and rock physics. *Science* 270:458–61.

Kellogg, L. H., and Turcotte, D. L. 1986–87. Homogeneization of the mantle by convective mixing and diffusion. *Earth Planet. Sci. Lett.* 81:371–78.

Kellogg, L. H., Hager, B. H., and van der Hilst, R. D. 1999. Compositional stratification in the deep mantle. *Science* 283:1881–84.

Kendall, J-M., and Silver, P. G. 1996. Constraints from seismic anisotropy on the nature of the lowermost mantle. *Nature* 381:409–12.

Kennett, B. L. N., Widiyantoro, S., and van der Hilst, R. D. 1998. Joint seismic tomography for bulk sound and shear wave speed in the Earth's mantle. *J. Geophys. Res.* 103:12469–93.

Kikuchi, M., and Kanamori, H. 1994. The mechanisms of the deep Bolivia earthquake of June 9, 1994. *Geophys. Res. Lett.* 22:2341–44.

Kimura, K., Lewis, R. S., and Anders, E. 1974. Distribution of gold and rhenium between nickel-iron and silicate melts: Implications for the abundances of the siderophile elements in the Earth and Moon. *Geochim. Cosmochim. Acta* 38:683–701.

Kirby, S. H. 1987. Localized polymorphic phase transformations in high-pressure faults and applications to the physical mechanisms of deep earthquakes. *J. Geophys. Res.* 92:13789–800.

Kirby, S. H., Durham, W. B., and Stern, L. 1991. Mantle phase changes and deep-earthquake faulting in subducting slabs. *Science* 252:216–25.

Kirby, S. H., Engdahl, E. R., and Denlinger, R. 1996. Intermediate-depth intraslab earthquakes and arc volcanism as physical expressions of crustal and uppermost mantle metamorphism in subducting slabs (overview). In *Subduction Top to Bottom,* edited by G. E. Bebout, D. W. Scholl, S. H. Kirby, and J. P Platt, pp. 195–214. Washington, D.C.: Amer. Geophys. Union.

Kirby, S. H., Stein, S., Okal, E. A., and Rubie, D. C. 1996. Deep earthquakes and metastable mantle phase transformations in subducting oceanic lithosphere. *Rev. Geophys.* 34:261–306.

Kohlstedt, D. L. 1992. Structure, rheology, and permeability of partially molten rocks at low-melt fraction. In *Mantle Flow and Melt Generation at Mid-Ocean*

Ridges, edited by J. Phipps Morgan, D. K. Blackman, and J. M. Sinton, pp. 103–121. Washington, D.C.: Amer. Geophys. Union.

Kohlstedt, D. L., Evans, B., and Mackwell, S. J. 1995. Strength of the lithosphere: Constraints imposed by laboratory experiments. *J. Geophys. Res.* 100:17587–602.

Kohlstedt, D. L., Keppler, H., and Rubie, D. C. 1996. Solubility of H_2O in the α, β, and γ phases of $(Mg,Fe)_2SiO_4$. *Contrib. Mineral. Petrol.* 123:345–57.

Koper, K. D., Wiens, D. A., Dorman, L. M., Hildebrand, J. A., and Webb, S. C. 1998. Modeling the Tonga slab: Can travel time data resolve a metastable olivine wedge? *J. Geophys. Res.* 103:30079–100.

Kuang, W., and Bloxham, J. 1997. An Earth-like numerical dynamo model. *Nature* 389:371–74.

Kubo, A., and Hiramatsu, Y. 1998. On the presence of seismic anisotropy in the asthenosphere beneath continents and its dependence on plate velocity: Significance of reference frame selection. *PAGEOPH* 151:281–303.

Kubo, T., Kato, T., Ohtani, E., Urakawa, S., Suzuki, A., Funakoshi, K., Utsumi, W., and Fujino, K. 2000. Formation of metastable assemblages and mechanisms of the grain-size reduction in the postspinel transformation of Mg_2GeO_4. *Geophys. Res. Lett.* 27:807–10.

Kubo, T., Ohtani, E., Kato, T., Shinmei, T., and Fujino, K. 1998. Effects of water on the α-β transformation in San Carlos olivine. *Science* 281:85–87.

Larmor, J. 1919. Possible rotational origin of magnetic fields of Sun and Earth. *Elec. Rev.* 85:412.

Laske, G., and Masters, G. 1999. Limits on differential rotation of the inner core from an analysis of the Earth's free oscillations. *Nature* 402:66–69.

Lay, T., Williams, Q., and Garnero, E. 1998. The core-mantle boundary layer and deep Earth dynamics. *Nature* 392:461–68.

Lee, D-C., and Halliday, A. N. 1995. Hafnium-tungsten chronometry and the timing of terrestrial core formation. *Nature* 378:771–74.

Li, J., and Agee, C. B. 2001. The effects of pressure, temperature, oxygen fugacity on partitioning of nickel and cobalt between liquid Fe-Ni-S alloy and liquid silicate: Implications for the Earth's core formation. *Geochim. Cosmochim. Acta* 65:1821–32.

Mackwell, S. J., and Kohlstedt, D. L. 1990. Diffusion of hydrogen in olivine: Implications for water in the mantle. *J. Geophys. Res.* 95:5079–88.

Mackwell, S. J., Kohlstedt, D. L., and Paterson, M. S. 1985. The role of water in the deformation of olivine single crystals. *J. Geophys. Res.* 90:11319–33.

Manga, M. 1996. Mixing of heterogeneities in the mantle: Effect of viscosity differences. *Geophys. Res. Lett.* 23:403–6.

Mao, H.-K., and Hemley, R. J. 1998. New windows on the earth's deep interior. *In Ultrahigh-Pressure Mineralogy: Physics and Chemistry of the Earth's Interior,* edited by R. J. Hemley, pp. 1–32. Washington, D.C.: Mineralogical Society of America.

Maruyama, S. 1994. Plume tectonics. *J. Geol. Soc. Japan* 100:24–49.

Masters, G., Jordan, T. H., Silver, P. G., and Gilbert, F. 1982. Aspherical Earth structure from fundamental spheroidal-mode data. *Nature* 298:609–13.

Masters, G., Laske, G., Bolton, H., and Dziewonski, A. M. 2000. The relative behavior of shear velocity, bulk sound speed, and compressional velocity in the mantle: Implications for chemical and thermal structure. In *Earth's Deep Interior,* edited by S. Karato, A. M. Forte, R. C. Liebermann, G. Maters, and L. Stixrude, pp. 63–87. Washington, D.C.: Amer. Geophys. Union.

McGovern, P. J., and Schubert, G. 1989. Thermal evolution of the Earth: Effects of volatile exchange between atmosphere and interior. *Earth Planet. Sci. Inter.* 96:27–37.

McKenzie, D. P. 1967. Some remarks on heat flow and gravity anomalies. *J. Geophys. Res.* 72:6261–73.

———. 1969. Speculations on the consequences and causes of plate motions. *Geophys. J. R. Astr. Soc.* 18:1–32.

McNamara, A. K., Karato, S., and van Keken, P. E. 2001. Localization of dislocation creep in the lower mantle: Implications for the origin of seismic anisotropy. *Earth Planet. Sci. Lett.* 191:85–99.

McNamara, A. K., van Keken, P. E., and Karato, S. 2002. Development of anisotropic structure by solid-state convection in the Earth's lower mantle. *Nature* 416:310–14.

Meade, C., and Jeanloz, R. 1991. Deep-focus earthquakes and recycling of water into the Earth's mantle. *Science* 252:68–72.

Mei, S., and Kohlstedt, D. L. 2000. Influence of water on plastic deformation of olivine aggregates, pt. 1: Diffusion creep regime. *J. Geophys. Res.* 105:21457–69.

Meissner, R., and Mooney, W. 1998. Weakness of the lower continental crust: A condition for delamination, uplift, and escape. *Tectonophysics* 296:47–60.

Merrill, R. T., McElhinney, M. E., and McFadden, P. L. 1998. *The Magnetic Field of the Earth.* New York: Academic Press.

Minster, J. B., and Anderson, D. L. 1981. A model for dislocation-controlled rheology for the mantle. *Phil. Trans. Roy. Soc. London* 299:319–56.

Mitrovica, J. X., and Peltier, W. R. 1991. Radial resolution in the inference of mantle viscosity from observations of glacial isostatic adjustment. In *Glacial Isostasy, Sea-Level, and Mantle Rheology,* edited by R. Sabadini, K. Lambeck, E. Boschi, pp. 63–78. Dordrecht: Kluwer Academic.

Mizutani, H., and Kanamori, H. 1964. Variation in elastic wave velocity and attenuative property near the melting temperature. *J. Phys. Earth* 12:43–49.

Montagner, J-P., and Guillot, L. 2000. Seismic anisotropy in the earth's mantle. In *Problems in Geophysics for the New Millennium,* edited by E. Boschi, G. Ekström, and A. Morelli, pp. 213–53. Rome: Editrice Compositiori.

Montagner, J-P., and Kennett, B. L. N. 1996. How to reconcile body-wave and normal-mode reference Earth models. *Geophys. J. Int.* 125:229–48.

Montagner, J-P., and Ritsema, J. 2001. Interactions between ridges and plumes. *Science* 294:1472–73.

Montagner, J-P., and Tanimoto, T. 1990. Global anisotropy in the upper mantle inferred from regionalization of the phase velocities. *J. Geophys. Res.* 95:4707–4819.

———. 1991. Global upper mantle tomography of seismic velocities and anisotropies. *J. Geophys. Res.* 96:20337–51.

Mooney, W. D., Laske, G., and Masters, G. 1998. CRUST 5.1: A global crystal model at 5° × 5°. *J. Geophys. Res.,* 103:727–47.

Morelli, A., Dziewonski, A. M., and Woodhouse, J. H. 1986. Anisotropy of the inner core inferred from PKIKP travel-times. *Geophys. Res. Lett.* 13:1545–48.

Mosenfelder, J. D., Marton, F. C., Ross, C. R., II., Kerschhofer, L., and Rubie, D. C. 2001. Experimental constraints on the depth of olivine metastability in subducting lithosphere. *Phys. Earth Planet. Inter.* 127:165–80.

Murthy, V. R. 1991. Early differentiation of the Earth and the problem of mantle siderophile elements: A new approach. *Science* 253:303–6.

Murthy, V. R., and Karato, S. 1997. Core formation and chemical equilibrium in the Earth, pt. 2: Chemical consequences for the mantle and core. *Phys. Earth Planet. Inter.* 100:61–79.

Nakada, M., and Lambeck, K. 1989. Late Pleistocene and Holocene sea-level change in the Australian region and mantle rheology. *Geophys. J. Int.* 96:495–517.

Nataf, H-C., Nakanishi, I., and Anderson, D. L. 1986. Measurements of mantle wave velocities and inversion for lateral heterogeneity and anisotropy. *J. Geophys. Res.* 91:7261–307.

Navrotsky, A. 1980. Lower mantle phase transitions may generally have negative pressure-temperature slopes. *Geophys. Res. Lett.* 7:709–711.

Nicolas, A., and Christensen, N. I. 1987. Formation of anisotropy in upper mantle peridotite—A review. In *Composition, Structure, and Dynamics of the Lithosphere-Asthenosphere Structure,* edited by K. Fuchs and C. Froidevaux, pp. 111–23. Washington, D.C.: Amer. Geophys. Union.

Nolet, G. 1987. Waveform tomography, In *Seismic Tomography,* edited by G. Nolet, pp. 201–22. Dordrecht, Holland: D. Reidel Publishing Co.

Obata, M., and Karato, S. 1995. Ultramafic pseudotachylyte from Balmuccia peridotite, Ivrea-Verbano zone, northern Italy. *Tectonophysics* 242:313–28.

Ogawa, M. 1987. Shear instability in a viscoelastic material as the cause for deep-focus earthquakes. *J. Geophys. Res.* 92:13801–10.

Ohtani, E. 1985. The primordial magma ocean and its implications for stratification of the mantle. *Earth Planet. Sci. Lett.* 78:70–80.

Okuchi, T. 1997. Hydrogen partitioning into molten iron at high pressure: Implications for Earth's core. *Science* 278:1781–84.

Olson, P., Christensen, U. R., and Glatzmaier, G. A. 1999. Numerical modeling of the geodynamo—Mechanisms of field generation and equilibration. *J. Geophys. Res.* 104:10383–404.

O'Neill, B., and Jeanloz, R. 1994. $MgSiO_3$-$FeSiO_3$-Al_2O_3 in the Earth's lower mantle. *J. Geophys. Res.* 99:19901–15.

Panasyuk, S. V., and Hager, B. H. 1998. A model of transformational superplasticity in the upper mantle. *Geophys. J. Int.* 133:741–55.

Park, J., and Levin, V. 2002. Seismic anisotropy: tracing plate dynamics in the mantle. *Science* 296:485–89.

Paterson, M. S. 1983. Creep in transforming polycrystalline materials. *Mech. Mater.* 2:103–9.

———. 1989. The interaction of water with quartz and its influence in dislocation flow—An overview. In *Rheology of Solids and of the Earth,* edited by S. Karato and M. Toriumi, pp. 107–42. Oxford: Oxford University Press.

Peacock, S. 2001. Are the lower planes of double seismic zones caused by serpentine dehydration in subducting oceanic mantle? *Geology* 29:299–302.

Peltier, W. R. 1989. Mantle viscosity. In *Mantle Convection,* edited by W. R. Peltier, pp. 389–478. New York: Gordon & Breach.

———. 1998. Postglacial variation in the level of the sea—Implications for climate dynamics and solid-earth geophysics. *Rev. Geophys.* 36:603–89.

Poirier, J-P. 1994. Light elements in the Earth's outer core: A critical review. *Phys. Earth Planet. Inter.* 85:319–37.

Pollack, H. N. 1986. Cratonization and thermal evolution of the mantle. *Earth Planet. Sci. Lett.* 80:175–82.

Raitt, R. W., Shor, G. G., Francis, T. J. G., and Morris, G. B. 1969. Anisotropy of the Pacific upper mantle. *J. Geophys. Res.* 74:3095–109.

Raleigh, C. B., and Paterson, M. S. 1965. Experimental deformation of serpentinite and its tectonic implications. *J. Geophys. Res.* 70:3965–85.

Regan, J., and Anderson, D. L. 1984. Anisotropic models of the upper mantle. *Phys. Earth Planet. Sci.* 35:227–63.

Richards, M. A., and Davies, G. F. 1989. On the separation of relatively buoyant components from subducted lithosphere. *Geophys. Res. Lett.* 16:831–34.

Richards, M. A., and Hager, B. H. 1984. Geoid anomalies in a dynamic Earth. *J. Geophys. Res.* 89:5987–6002.

Riedel, M. R., and Karato, S. 1997. Grain-size evolution in subducted oceanic lithosphere associated with the olivine—Spinel transformation and its effects on rheological weakening. *Earth Planet. Sci. Lett.* 148:27–43.

Ringwood, A. E. 1966. Chemical evolution of the terrestrial planets. *Geochem. Cosmochim. Acta* 30:41–104.

———. 1967. The pyroxene-garnet transformation in the Earth's mantle. *Earth Planet. Sci. Lett.* 2:255–63.

———. 1975. *Composition and Petrology of the Earth's Mantle.* New York: McGraw-Hill.

———. 1982. Phase transformations and differentiation in subducted lithosphere—Implications for mantle dynamics, basalt petrogenesis, and crustal evolution. *J. Geol.* 90:611–43.

———. 1991. Phase transformations and their bearings on the constitution and dynamics of the mantle. *Geochim. Cosmochim. Acta* 55:2083–110.

————. 1994. The role of transition zone and 660 km discontinuity in mantle dynamics. *Phys. Earth Planet. Inter.* 86:5–24.

Robertson, G. S., and Woodhouse, J. H. 1996a. Constraints on lower-mantle physical properties from seismology and mineral physics. *Earth Planet. Sci. Lett.* 143:197–205.

————. 1996b. Ratio of S to P velocity heterogeneity in the lower mantle. *J. Geophys. Res.* 101:20041–52.

Romanowicz, B., and Durek, J. J. 2000. Seismological constraints on attenuation in the Earth. In *Earth's Deep Interior,* edited by S. Karato, A. M. Forte, R. C. Liebermann, G. Maters, and L. Stixrude, pp. 161–79. Washington, D.C.: Amer. Geophys. Union.

Rubie, D. C., Tsuchida, Y. Yagi, T., Utsumi, W., Kikegawa, T., Shimamura, O., and Brearley, A. J. 1990. An in situ X-ray diffraction study of the kinetics of the Ni_2SiO_4 olivine-spinel transformation. *J. Geophys. Res.* 95:15829–44.

Rudnick, R. L., and Fountain, D. M. 1995. Nature and composition of the continental crust: A lower crustal perspective. *Rev. Geophys.* 33:267–309.

Runcorn, S. K. 1983. Lunar magnetism, polar displacement, and primeval satellites in the Earth-Moon system. *Nature* 304:589–96.

Sammis, C. G., and Dein, J. L. 1974. On the possibility of transformational superplasticity in the Earth's mantle. *J. Geophys. Res.* 79:2961–65.

Sato, H., Sacks, I. S., Murase, T., Muncill, G., and Fukuyama, H. 1989. Q_p-melting temperature relation in peridotite at high pressure and temperature: Attenuation mechanism and implications for the mechanical properties of the upper mantle. *J. Geophys. Res.* 94:10647–61.

Savage, M. K., and Silver, P. G. 1993. Mantle deformation and tectonics: Constraints from seismic anisotropy in the western United States. *Phys. Earth Planet. Inter.* 78:207–27.

Schubert, G. 1997. Inside the solid planets and moons. *Phys. World* (Oct.):45–49.

Sclater, J. G., Parsons, B., and Jaupart, C. 1981. Oceans and continents: Similarities and differences in the mechanisms of heat loss. *J. Geophys. Res.* 86:11535–52.

Seno,T., and Yamanaka, Y. 1996. Double seismic zones, deep compressional trench-outer rise events and superplumes. In *Subduction from Top to Bottom,* edited by G. E. Bebout, D. W. Scholl, S. H. Kirby, and J. P. Platt, pp. 347–55. Washington, D.C.: Amer. Geophys. Union.

Shankland, T. J., O'Connell, R. J., and Waff, H. S. 1981. Geophysical constraints on partial melt in the upper mantle. *Rev. Geophys. Space Phys.* 19:394–406.

Shearer, P. M., and Masters, G. 1992. Global mapping of topography on the 660-km discontinuity. *Nature* 355:791–96.

Silver, P. G. 1996. Seismic anisotropy beneath the continents—Probing the depth of geology. *Ann. Rev. Earth Planet. Sci.* 24:385–432.

Silver, P. G., Beck, S. L., Wallace, T. C., Meade, C., Myers, S. C., James, D. E., and Kuehnel, R. 1995. Rupture characteristics of the Bolivia earthquake of 9 June 1994 and the mechanism of deep-focus earthquakes. *Science* 268:69–71.

Silver, P. G., Carlson, R. W., and Olson, P. 1988. Deep slabs, geochemical heterogeneity, and the large-scale structure of mantle convection: Investigation of an enduring paradox. *Ann. Rev. Earth Planet. Sci.* 16:477–541.

Sleep, N. H. 1988. Gradual entrainment of a chemical layer at the base of the mantle by overlying convection. *Geophys. J.* 95:437–47.

———. 1990. Hotspots and mantle plumes—Some phenomenology. *J. Geophys. Res.* 95:6715–36.

Solomatov, V. S. 1996. Can hotter mantle have a larger viscosity? *Geophys. Res. Lett.* 23:937–40.

Song, X. 1997. Anisotropy of the Earth's inner core. *Rev. Geophys.* 35:297–313.

Song, X., and Richards, P. G. 1996. Seismological evidence for differential rotation of the Earth's inner core. *Nature* 382:221–24.

Souriau, A. 1998. Earth's inner core: Is rotation real? *Science* 281:55–56.

Spetzler, H. A., and Anderson, D. L. 1968. The effect of temperature and partial melting on velocity and attenuation in a simple binary mixture. *J. Geophys. Res.* 73:6051–60.

Steinle-Neumann, G., Stixrude, L., Cohen, R. E., and Gülseren, O. 2001. Elasticity of iron at the temperature of the Earth's inner core. *Nature* 413:57–60.

Stevenson, D. J. 1981. Models of the Earth's core. *Science* 214:611–19.

———. 1990. Fluid dynamics of core formation. In *The Origin of the Earth*, edited by H. E. Newsom and J. H. Jones, pp. 231–49. Oxford: Oxford University Press.

Stevenson, D. J, Spohn, T., and Schubert, G. 1983. Magnetism and thermal evolution of the terrestrial planets. *Icarus.* 54:466–89.

Stixrude, L., and Cohen, R. E. 1995. High-pressure elasticity of iron and anisotropy of earth's inner core. *Science* 267:1972–75.

Stixrude, L., Hemley, R. J., Fei, Y., and Mao, H-K. 1992. Thermoelasticity of silicate perovskite and magnesiowüstite and the stratification of the Earth's mantle. *Science* 257:1099–1101.

Stocker, R. L., and Gordon, R. B. 1975. Velocity and internal friction in partial-melts. *J. Geophys. Res.* 80:4828–36.

Su, W-J., Dziewonski, A. M., and Jeanloz, R. 1996. Planet within a planet: Rotation of the inner core of the Earth. *Science* 274:1883–87.

Sumita, I., Yoshida, S., Kumazawa, M., and Hamano, Y. 1996. A model for sedimentary compaction of a viscous medium and its application to inner-core growth. *Geophys. J. Int.* 124:502–24.

Sung, C. M., and Burns, R. G. 1976. Kinetics of high-pressure phase transformations—Implications to the evolution of the olivine-spinel transition in the downgoing lithosphere and its consequences on the dynamics of the mantle. *Tectonophysics* 31:1–31.

Tackley, P. J., Stevenson, D. J., Glatzmaier, G. A., and Schubert, G. 1993. Effects of an endothermic phase transition at 670 km depth on spherical mantle convection. *Nature* 361:699–704.

Takahashi, E., Nakajima, K., and Wright, T. L. 1998. Origin of the Columbia River basalts: Melting model of a heterogeneous plume head. *Earth Planet. Sci. Lett.* 162:63–80.

Tanimoto, T., and Anderson, D. L. 1984. Mapping mantle convection. *Geophys. Res. Lett.* 11:287–90.

Tonks, W. B., and Melosh, H. J. 1990. The physics of crystal settling and suspension in a turbulent magma ocean. In *Origin of the Earth,* edited by H. E. Newsom and J. H. Jones, pp. 151–74. Oxford: Oxford University Press.

Turekian, K., and Clark, S. P., Jr. 1969. Inhomogeneous accumulation of the Earth from the primitive solar nebula. *Earth Planet. Sci. Lett.* 6:346–48.

van der Hilst, R. D., Engdahl, E. R., Spakman, W., and Nolet, G. 1991. Tomographic imaging of subducted lithosphere below northwest Pacific island arcs. *Nature* 353:37–43.

van der Hilst, R. D., and H. Karason 1999. Compositional heterogeneity in the bottom 1,000 kilometers of Earth's mantle: Toward a hybrid convection model. *Science* 283:1885–88.

van der Hilst, R. D., and Seno, T. 1993. Effects of relative plate motion on the deep structure and penetration depth of slabs below the Izu-Bonin and Mariana arcs. *Earth Planet. Sci. Lett.* 120:395–407.

van der Hilst, R. D., Widiyantoro, S., and Engdahl, E. R. 1997. Evidence for deep mantle circulation from global tomography. *Nature* 386:578–84.

van Keken, P. E., and Zhong, S. 1999. Mixing in a 3-D spherical model of present-day mantle convection. *Earth Planet. Sci. Lett.* 171:533–47.

Vaughan, P. J., and Coe, R. S. 1981. Creep mechanisms in Mg_2GeO_4: Effects of a phase transition. *J.Geophys. Res.* 86:389-404.

Vinnik, L. P., Green, R. W. E., and Nicolaysen, L. O. 1995. Recent deformation of the deep continental root beneath southern Africa. *Nature* 375:50–52.

Wadati, K. 1927. Existence and study of deep earthquakes. *J. Meteorol. Soc. Jpn.,* ser. 2, 5:119–45.

Walker, K. T., Bokelmann, G. H. R., and Klemperer, S. L. 2001. Shear-wave splitting to test mantle deformation models around Hawaii. *Geophys. Res. Lett.* 28:4319–22.

Weidner, D. J. 1998. Rheological studies at high pressure. In *Ultrahigh-Pressure Mineralogy,* edited by R. J. Hemley, pp. 493–524. Washington, D.C.: Mineral. Soc. Amer.

Wiens, D. A., and Gilbert, H. J. 1996. Effect of slab temperature on deep earthquake aftershock productivity and magnitude-frequency relations. *Nature* 384:153–56.

Wiens, D. A., McGuire, J. J., Shore, P. J., Bevis, M. G., Draunidalo, K., Prasad, G., and Helu, S. P. 1994. A deep earthquake aftershock sequence and implications for the rupture mechanism of deep earthquakes. *Nature* 372:540–43.

Wiens, D. A., and Snider, N. O. 2001. Repeating deep earthquakes: Evidence for fault reactivation at great depth. *Science* 293:1463–66.

Wetherill, G. W. 1990. Formation of the Earth. *Ann. Rev. Earth Planet. Sci.* 18:205–56.

Wolfe, C. J., Bjarnason, I. Th., VanDecar, J. C., and Solomon, S. C. 1997. Seismic structure of the Iceland mantle plume. *Nature* 385:245–47.

Wolfe, C. J., and Solomon, S. C. 1998. Shear-wave splitting and implications for mantle flow beneath the MELT region of the East Pacific Rise. *Science* 280:1230–32.

Woodhouse, J. H., and Dziewonski, A. M. 1984. Mapping the upper mantle—Three-dimensional modeling of Earth structure by inversion of seismic waveforms. *J. Geophys. Res.* 89:5953–86.

Woodhouse, J. H., Giardini, D., and Li, X. D. 1986. Evidence for inner core anisotropy from free oscillations. *Geophys. Res. Lett.* 31:1549–52.

Wookey, J., Kendall, J-M., and Barruol, G. 2002. Evidence of mid-mantle deformation from seismic anisotropy. *Nature* 415:778–80.

Wong, T-f., Ko, S., and Olgaard, D. L., 1997. Generation and maintenance of pore pressure excess in a dehydrating system, pt. 2: Theoretical analysis. *J. Geophys. Res.* 102:841–52.

Yamazaki, D., and Karato, S. 2001a. High-pressure rotational deformation apparatus to 15 GPa. *Rev. Sci. Instrum.* 72:4207–11.

———. 2001b. Some mineral physics constraints on the rheology of Earth's lower mantle. *Amer. Mineral.* 86:385–91.

———. 2002. Fabric development in (Mg,Fe)O during large strain, shear deformation: Implications for seismic anisotropy in Earth's lower mantle. *Phys. Earth Planet. Inter.* 131:251–67.

Yan, H. 1992. "Dislocation Recovery in Olivine." MSc. thesis. University of Minnesota.

Yoshida, S., Sumita, I., and Kumazawa, M. 1996. Growth model of the inner core coupled with the outer core dynamics and the resulting elastic anisotropy. *J. Geophys. Res.* 101:28085–103.

Yoshii, T. 1975. Regionality of group velocities of Rayleigh waves in the Pacific and thickening of the plate. *Earth Planet. Sci. Lett.* 25:305–12.

Yoshii, T., Kono, Y., and Ito, K. 1976. Thickening of the oceanic lithosphere. In *The Geophysics of the Pacific Ocean Basin and Its Margin*, pp. 424–30. Washington, D.C.: Amer. Geophys. Union.

Zhang, Y-S., and Tanimoto, T. 1991. Global Love wave phase velocity variation and its significance to plate tectonics. *Phys. Earth. Planet. Inter.* 66:160–202.

———. 1993. High-resolution global upper mantle structure and plate tectonics. *J. Geophys. Res.* 98:9793–823.

Zhao, D. 2001. Seismic structure and origin of hotspots and mantle plumes. *Earth Planet. Sci. Lett.* 192:251–65.

Zimmerman, M. E., Zhang, S., Kohlstedt, D. L., and Karato, S. 1999. Melt distribution in mantle rocks deformed in shear. *Geophys. Res. Lett.* 26:1505–08.

INDEX

660-km boundary, 121; crust separation and, 146–52; grain size and, 139–41; Nusselt number and, 128; Peierls stress and, 134–35; phase transformations and, 122–29, 139–41; rheological effects and, 129–46; scaling laws and, 137–38; slab thermal parameter and, 132–33

achondrite, 188
adiabatic processes: boundary layer theory and, 32–34; Bullen parameter and, 16; compression and, 15–16; hardening and, 64–66; instability and, 174
aftershocks, 162, 178, 180
Agee, Carl, 5, 188
Aki, Keiiti, 73, 98
Akimoto, Syun-iti, 17
Alfvén, Hannes, 199–200
Allegre, Claude, 127
Anderson, Don, 5, 11–12, 28; anelasticity and, 56; heterogeneity studies and, 73; partial melting and, 52–53
Anderson, Orson, 88
Anderson-Grüneisen parameter, 88
Ando, Masataka, 98
anelasticity, 12, 34, 37, 117; asthenosphere and, 52, 55–57; attenuation and, 89; heterogeneity and, 82, 88–91. *See also* elasticity
anharmonicity, 35, 52; heterogeneity and, 82–87
anisotropy, vii–viii, 72, 117–18; azimuthal, 73, 95–97; core and, 185, 204–12; creep and, 110–11; deformation and, 107, 109–10, 112–16; degree of, 105; description of, 94–96; dictionary for, 117; heterogeneity and, 98; iron and, 115–16; layered structure and, 103–6; lithosphere and, 101; magnesium and, 115–16; mineral physics and,

103, 106–16; oceans and, 97–99; olivine and, 110–15; polarization, 95, 97, 102–3, 113–14; preferred orientation and, 103, 106–16; seismology and, 96–103; strength of, 106; transverse isotropy and, 105–6; types of, 95–96; water and, 112–15
anti-continents, vii
applied mathematics. *See* mathematics
Archean regions, 47
asthenosphere, 48, 50; anelasticity and, 55–57; body waves and, 46–47; boundary conditions and, 51; description of, 44; hardening and, 64–66; new model for, 64–67; oceans and, 45–47; partial melting and, 51–55; structure of, 44–47; surface waves and, 46
Aurnou, Jon, 212
Avrami length, 139–40
Avrami time, 139

basalt, 2, 6, 119; convection and, 120; MORB, 3, 120, 152; oceanic separation and, 146, 152; OIB, 146, 152
bathymetry, 48, 50
Becker, Thorsten, 154–55
Benioff, Hugo, 158–59
Bercovici, David, 128
Bergman, Michael, 208, 210
Birch, Francis, 2, 15, 52, 85
Blacic, Jim, 58, 60
blobs, 153–57
Bloxham, Jerry, 197
body waves, 46–47, 76
boundary layer theory, 32–34
Bridgman, Percy, 15, 168–69
brittle deformation, 163–66
Buffett, Bruce, 184, 210–11
bulk modulus, 9, 28
Bullard, Edward, 196

Bullen, Keith, 16
Bullen parameter, 16–17, 19
buoyancy, 35
Byerlee, Jim, 165

Callisto, 193
carbonaceous chondrite, 3–5
chemistry. *See* geochemistry
Christensen, Nick, 110
Christensen, Ulrich, 124–29, 146
Clapeyron slope, 21–22, 120; convection
    and, 125; entropy and, 23–24
Clark, Sydney, 187
cobalt, 186, 188
Cohen, Ron, 207
cold tongue, 133
collision zones, 71
compatible elements, 6–7
composition: chemical, 26; convection and,
    120; core, 14, 182–85; geochemical
    models and, 2–7; geophysical models
    and, 7–14; heterogeneity and, 25, 27–
    29, 73–93; homogeneity and, 25, 27–
    29; inferring of, 1; layered structure and,
    15–29; phase transformations and, 15–
    25, 122–29; reservoirs and, 120; seismic
    wave attenuation and, 34–43; subduc-
    tion and, 119–52 (*see also* subduction).
    See also material properties
compression, 9; 660-km boundary and,
    122; anisotropy and, 94–117; displace-
    ment and, 104–6; earthquakes and, 158
    (*see also* earthquakes); faults and, 161–
    62, 164–76; heterogeneity and, 80;
    Peierls stress and, 134–35; phase trans-
    formations and, 15–25; yield stress and,
    134–35
convection, viii, 31–32; 660-km boundary
    and, 122–29; chemical differentiation
    and, 120; Clapeyron slope and, 125;
    grain size and, 133–44; heat effects and,
    126; magnetism and, 199–201; Nusselt
    number and, 126, 128; phase transfor-
    mation and, 15–25, 122–29; Rayleigh
    number and, 123–24, 126; rheological
    effects and, 129–38, 141–46; subduc-
    tion and, 119–52 (*see also* subduction);

trenches and, 130–31. *See also* thermal
    structure
core: anisotropy and, 185, 204–12; com-
    position of, 14, 182–85; Coriolis force
    and, 198–99; dendrite and, 184; density
    and, 14; dynamics of inner, 203–12; dy-
    namics of outer, 194–203; dynamo the-
    ory and, 196–203; Ekman boundary
    layer, 213–14; elasticity and, 185; en-
    ergy sources of, 194–96; evolution of,
    185–94; fluid dynamics and, 199–201,
    206–12; free oscillation and, 204; geo-
    magnetism and, 15; gravity and, 189–
    90; heterogeneity and, 210–11; homo-
    geneity of, 25, 27; Lorentz force and,
    198–99; magnetism and, 182, 193–94;
    phase transitions and, 15–25; radioac-
    tive decay and, 185; sediments and,
    184; seismology of, 203–5, 212–14;
    siderophile elements and, 186–87; struc-
    ture of, 182–94; super-rotation and,
    204–5, 212; Taylor column and, 199,
    201, 210; Taylor number and, 199; tem-
    perature and, 207; thermal structure of,
    29–34; toroidal field of, 201–2, 212;
    westward drift and, 194; white inclu-
    sions and, 187–88; zeroth-order approx-
    imation and, 14
Coriolis force, 198–99
Coulomb-Navier's law, 164–65
Creager, Ken, 205
creep, 10, 133; blobs in matrix, 153–57;
    diffusion and, 59–61, 137, 140; disloca-
    tion and, 107–8, 110–11, 137; grain
    size and, 135–36
crust: depleted, 152; description of, 14; en-
    riched, 152; lithosphere separation and,
    146–52; mixing and, 152–57; subduc-
    tion and, 146–52; tectosphere and, 68–
    71; undepleted, 152
crystallization: core and, 206–12; defor-
    mation and, 61–64, 107–8; dislocations
    and, 59–60; preferred orientation and,
    103, 106–16. *See also* anisotropy

Davies, Geoff, 129, 148
deformation, 62–64; anisotropy and, 72–

73, 103–18; blobs in matrix, 153–57; brittle-ductile transition and, 163–66; convection and, 129–38, 141–46; creep and, 110, 133; crystallization and, 107–8; diffusion creep and, 59–61; dislocation creep and, 107–8, 110–11; ductile, 158, 161–76; faults and, 161–62, 164–76; magnetism and, 202; Peierls stress and, 134–35; preferred orientation and, 103, 106–16; rate of, 59–60; thermal runaway and, 174–79; yield stress and, 134–35. *See also* earthquakes

dehydration, 168, 180–81

delamination, 71

dendrite, 184

density, 2, 39; Bullen parameter and, 16–17; chemical composition and, 91–93; Clapeyron slope and, 125; compositional homogeneity and, 25, 27–29; convection and, 129–38, 141–46; core and, 14, 182–85; Earth and, 7; earthquakes and, 8–9; free oscillation and, 77; heterogeneity and, 73–93; inertia and, 7–8; isostasy and, 48–50; lattice defects and, 58–60; lithosphere and, 48–51; oceanic separation and, 146–52; phase transformations and, 15–25, 122–29 (*see also* phase transformations); stress origins and, 165, 167; subduction and, 119–52 (*see also* subduction); synchrotrons and, 26–27; of water, 50

depleted regions, 153–57

diamond anvil cell (DAC), 18–19

diamonds, 25, 136, 138

diffusion creep, 59–61, 137, 140

dihedral angle, 53–55

dislocation, 59–60; creep, 107–8, 110, 137

disorder, 23

dispersion, 10

D″ layer, 31, 74, 102, 184

Durham, Bill, 117

dynamic topography, 39–40

dynamo theory, 196–203

Dziewonski, Adam, 11, 73

Earth: advances in study of, vii–ix; anisotropy and, 94–116; asthenosphere and, 44–48, 50–57, 64–67; chemical composition of, 25–29; core of, 182–214; Coriolis force and, 198–99; density of, 7; earthquakes and, 158 (*see also* earthquakes); first-order approximation and, 15–25; geochemical models of, 2–7; geophysical models of, 7–13; heat flux release of, 119; homogeneity and, 73–93; inferring composition of, 1; layered structure of, 15–29 (*see also* layers); lithosphere and, 44–50, 64–71; Lorentz force and, 198–99; phase transitions and, 15–25; rheological structure of, 34–43; second-order approximation and, 25–29; subduction and, 121–52; thermal structure of, 29–34; zeroth-order approximation and, 14

Earthquake Research Institute, 46

earthquakes: 660-km boundary and, 122; aftershocks and, 162, 178, 180; Bolivia, 162–63; brittle-ductile transition and, 163–66; characteristics of, 159–63; Colombia, 162; dehydration and, 168, 180–81; discovery of deep, 158–59; distribution of, 160–61; efficiency of, 162–63, 180; faults and, 161–62, 164–76; fracture propagation and, 172; free oscillation and, 11; hypocenter of, 8–9, 119; isolated, 161; isostasy and, 158; mechanisms of, 161–62, 165–78; mineral physics of, 171–74; nucleation and, 171–72; stress origins and, 165, 167; surface waves and, 9–11; temperature and, 158–59, 162, 164; thermal runaway and, 174–79; Wadati-Benioff zone and, 158–60; water and, 173–74, 180–81. *See also* seismology

eclogite, 2, 71

Ekman boundary layer, 213–14

elasticity, 117; anelasticity and, 52, 55–57 (*see also* anelasticity); anharmonicity and, 84–87; anisotropy and, 94–117; core and, 185, 206–12; deformation and, 35–36; earthquakes and, 163–66; heterogeneity and, 82–93; horizontal displacement and, 104–6; preferred orientation and, 103, 106–16; rheological

elasticity (*cont.*)
structure and, 34–43; vertical displacement and, 104–6
elastic waves. *See* seismology
electricity: conductivity, 44; dynamo theory and, 196–203; induction and, 197–98; Maxwell's equations and, 197; Ohm's law and, 197. *See also* magnetism
Elsasser, Walter, 196
endothermic transformation, 123
enriched regions, 153–57
enthalpy, 43
entropy, 23–24, 123
equations: anharmonicity, 86–87, 89; anistropic, 104–5; Avrami length, 139–40; Avrami time, 139; Bullen parameter, 16–17; Clapeyron slope, 21–23, cold tongue, 133; compression/shear wave velocity, 83–84; cooling time, 133; Coulomb-Navier's law, 165; diffusion creep, 137, 140; dislocation creep, 137; ductile deformation, 59–60, 165; dynamo theory, 197–98; elastic constants, 104; electromagnetic induction, 197; flexural rigidity, 144; Fourier's law, 50; Gibbs free energy, 22; Grüneisen parameter, 85, 87; lattice defects, 59–60; lithosphere, 48–50; Navier-Stokes, 198; Nusselt number, 128; ocean floor depth, 48–50; Peierls stress, 135; phase transformation, 22–23, 125; postglacial rebound, 38; Q-factor, 42; rate of nucleation, 139; Rayleigh number, 32; relaxation time, 38–39; Reynolds number, 198; rheological flow, 134–35, 137; scaling law, 137; seismic frequency, 56–57; seismic wave attenuation, 42; seismic wave velocity, 9, 56–57; slab thermal parameter, 132–33; subduction, 132–33; thermal expansion, 86, 88; thermal runaway, 175, 177–78; velocity/density anomalies, 84; water density, 50
Evans, Brian, 133
evolution: chemical, 6, 91; composition and, 120–21; of core, 185–94; oceanic lithosphere and, 47–51

Faraday, Michael, 196
faults, 161–62; brittle deformation and, 163–66; cracks and, 164, 166, 170; dehydration and, 168; formation of, 167–76; fracture propagation and, 72
first-order approximation, 15–25
flexural rigidity, 144
Forsyth, Don, 119
Forte, Alessandro, 37
Fourier's law, 50
free oscillation, 11; core and, 204; density and, 77; gravity and, 77 heterogeneity and, 73–74; oceanic separation and, 151–52
friction, 164–65
Fujii, Naoyuki, 54
Fukao, Yoshio, 76

Gaherty, Jim, 46–47, 148
Galileo spacecraft, 193
Ganymede, 193
Garnero, Ed, 102
garnet, 2, 146–52
Gauss, Friedrich, 8, 194
geochemistry: carbonaceous chondrite and, 3–5; core and, 25, 27, 185–94; defined, 1; differentiation theory and, 3; heterogeneity and, 91–93; mantle homogeneity and, 27–29; models of, 2–7; seismic tomography and, 91–93; subduction and, 119–52 (*see also* subduction)
geoid anomalies, 39–40
geomagnetism. *See* magnetism
geophysics, viii–ix; core and, 182–214; Coriolis force and, 198–99; defined, 1; dynamo theory and, 196–203; earthquakes and, 8–9, 158 (*see also* earthquakes); mixing and, 152–57; models of, 7–13; moment of inertia and, 7–8; precession and, 8; subduction and, 119–52 (*see also* subduction). *See also* Earth; mantle
GEOSCOPE, 117
glacial rebound, 38
Glatzmaier, Gary, 197, 201–2, 212
Goetze, Chris, 131, 133

scaling laws in, 137–38; siderophile elements and, 186–87; subduction and, 119–52 (*see also* subduction); thermal runaway and, 174–78; yield stress and, 134–35

mathematics, viii; bulk modulus, 9; Clapeyron slope, 21–23; Coulomb-Navier's law, 164–65; Gauss's theorem, 8; matrices, 154; Maxwell's equations, 197; Reynolds number, 198; Taylor number, 199. *See also* equations

matrices, 154

Maxwell, James, 196

Maxwell's equations, 197

McKenzie, Dan, 48, 50

Meade, Charlie, 173

megaliths, 148

Melosh, Jay, 5–6

melt pockets, 116

meteorites, 3, 188

mid-ocean ridge basalts (MORB), 3, 120, 152

mineral physics, vii–viii; anisotropy and, 103, 106–16; core and, 183–85; dictionary for, 117; earthquakes and, 171–74; heterogeneity and, 81–85, 87–93; lithophile elements and, 186, 188; magma ocean and, 5–6; pressure and, 64; siderophile elements and, 186–87; subduction and, 119–52; synchrotrons and, 26–27; tectosphere and, 68–71; temperature and, 64, 82–85, 87–91; water and, 46, 57–64

Minster, Bernard, 56

Mitrovica, Jerry, 37, 39

mixing, 152–57

Mizutani, Hitoshi, 52–53

modes of oscillation, 11

Montagner, Jean-Paul, 73, 77

Moon, 188

Murthy, Rama, 188

Nakada, Masao, 38

Nakanishi, Ichiro, 73

Navier-Stokes equation, 198

Newtonian fluid, 121

nickel, 186, 188

Nicolas, Adolfe, 110

Nolet, Guust, 74

nucleation: earthquakes and, 171–72; rate of, 136, 138–40, 171

numerical modeling, 146; asthenosphere 51–55, 64–67; boundary layer theory, 32–34; chemical structure of interior, 25–29; chondrite, 3–5; core formation, 185–94; dynamo theory, 196–203; Earth model, 13–29, 72–73, 75–76; earthquakes and, 159–76; first-order approximation, 15–25; geochemical, 2–7; geophysical, 7–13; hardening and, 64–66; heterogeneity and, 73–93; lithosphere and, 47–51, 64–67; magma ocean and, 5–6; partial melting, 44, 46–55; phase transformation, 15–25 (*see also* phase transformations); PREM, 11–12, 72; rheological structure, 34–43; second-order approximation, 25–29; thermal structure, 29–34; transformation-faulting, 169–73, 180; zeroth-order approximation, 14

Nusselt number, 126, 128

ocean island basalts (OIB), 146, 152

oceans: 660-km boundary and, 121–52; anisotropy and, 94–117; asthenosphere and, 45–47; body waves and, 46–47; convection and, 122–38, 141–46; depth equation for, 48–49; earthquakes and, 161; floor observations of, vii; hardening and, 64–66; homogeneity and, 76; lithosphere and, 45–51, 146–52; magma ocean and, 5–6; MORB and, 3, 120, 152; phase transformation and, 122–29; rheological effects and, 129–38, 141–46; subduction and, 119–52; tectosphere and, 68–71

Ogawa, Masaki, 174

Ohm's law, 197

Ohtani, Eiji, 5

olivine, 2, 5, 14; anisotropy and, 110–15; convection and, 123, 146; ductile deformation of, 61; faults and, 171; grain size and, 60–62; oceanic separation and, 147; preferred orientation and, 114;

olivine (*cont.*)
    pressure and, 19; tectosphere and, 69–70; water and, 58, 60–64
Olson, Peter, 201
orogenic belts, 99
oxygen, 63–64

partial melting, 44, 46; analog materials study and, 52–54; asthenosphere and, 51–55, 64–67; dihedral angle and, 53–55; hardening and, 64–66; heat flux release and, 119; oceanic lithosphere and, 47–51; subduction and, 119–52; tectosphere and, 68–71; temperature and, 82–85, 87–91; thermal runaway and, 174–78; water and, 57–58, 60–64
Paterson, Mervyn, 56, 60–61, 167
Peacock, Simon, 168
Peierls stress, 134–35
Peltier, Dick, 37, 39
peridotite, 2
perovskite, 19, 122–23
petrological approach, 2–3
phase transformations, 15, 43, 146; brittle-ductile transition and, 163–66; Bullen parameter and, 16–17, 19; Clapeyron slope and, 21–23; convection and, 122–29; density and, 19, 21; dimensional analysis of, 127–29; earthquakes and, 168–69 (*see also* earthquakes); faults and, 161–62, 164–76; grain size and, 133–44; kinetics of, 139–41; mantle and, 119–20; non-equilibrium, 25; subduction and, 119–52; temperature and, 21, 23–24; thermal runaway and, 174–78; transition zone and, 17, 19
plastic flow, 7
plate tectonics: 660-km boundary and, 121–52; asthenosphere and, 44–48, 50–57, 64–67; brittle-ductile transition and, 163–66; convection and, 129–38, 141–46; depth and, 80; development of, vii; earthquakes and, 157–63, 171 (*see also* earthquakes); heat flux release and, 119; heterogeneity and, 73–93; lithosphere and, 44–50, 64–71, 146–52; mixing influence and, 152–57; Peierls

stress and, 134–35; Rayleigh number and, 123–24; rheological effects and, 129–38, 141–46; subduction and, 119, 121–52; tectosphere and, 68–71; yield stress and, 134–35
plume shape, 128–29, 149
polarization anisotropy, 95, 97, 102–3, 113–14
Pollack, Henry, 70
precession, 8
preferred orientation, 103, 109, 112, 116; dislocation and, 107–8, 110; olivine and, 110–11, 114; recrystallization and, 106–7; water and, 113–15
Preliminary Reference Earth Model (PREM), 11–12, 72
pressure: Byerlee's law and, 165; Clapeyron slope, 21–23; convection and, 131, 133, 135; core and, 182–94, 206–12; depth and, 52; diamonds and, 18–19, 25; elasticity and, 35–36 (*see also* elasticity); faults and, 161–62, 164–76; Grüneisen parameter and, 85; hardening and, 64–66; heterogeneity and, 73–93; isostasy and, 48–50; mineral physics and, 64; phase transformations and, 15–25 (*see also* phase transformations); slab thermal parameter and, 132–33; subduction and, 119–52; synchrotrons and, 26–27; thermal expansion and, 88; viscosity variation and, 37–43; Wadati-Benioff zone and, 158–60. *See also* deformation; Rayleigh number
primary waves, 9–10
Proterozoic regions, 47
P-waves. *See* compression
pyrolite, 3
pyroxenes, 2, 5

Q-factor, 42
quantum theory, 117
quartz, 60; anisotropy and, 94; water and, 62–64

radioactive decay, 5, 185
Raleigh, Barry, 167
Rayleigh number: boundary layer theory